PRAISE FOR *NETWORK SOVEREIGNTY*

"Duarte shows that tribal ownership and use of information and communication technologies have the potential to deepen the meaning and experience of tribal sovereignty, serving as a means to undermine colonialism."
—**Andrew Needham**, author of *Power Lines: Phoenix and the Making of the Modern Southwest*

"The strength of *Network Sovereignty* is when the stories capture examples of sovereignty and technology in action."
—**Mark Trahant**, author of *The Last Great Battle of the Indian Wars: Henry M. Jackson, Forrest J. Gerard and the Campaign for the Self-Determination of America's Indian Tribes*

"In *Network Sovereignty*, Duarte looks at the psychological and philosophical implications of the colonization of Indigenous peoples in a technological age. She provides accessible and relevant examples of American Indians searching for ways to use new technologies to address very real social, cultural, and political challenges."
—**Ken Coates**, author of *#IdleNoMore: And the Remaking of Canada*

Charlotte Coté and Coll Thrush

Series Editors

NETWORK SOVEREIGNTY

Building the Internet across Indian Country

MARISA ELENA DUARTE

UNIVERSITY OF WASHINGTON PRESS
Seattle and London

www.tulalipcares.org

Network Sovereignty was made possible in part by a generous grant from the Tulalip Tribes Charitable Fund, which provides the opportunity for a sustainable and healthy community for all.

Printed and bound in the United States of America
Composed in Charter, typeface designed by Matthew Carter
21 20 19 18 17 5 4 3 2 1

University of Washington Press
www.washington.edu/uwpress

Library of Congress Cataloging-in-Publication Data
Names: Duarte, Marisa Elena, author.
Title: Network sovereignty : building the Internet across Indian Country / Marisa Elena Duarte.
Description: 1st edition. | Seattle : University of Washington, [2017] | Series: Indigenous confluences | Includes bibliographical references and index.
Identifiers: LCCN 2016047369| ISBN 9780295741819 (hardcover : alk. paper) | ISBN 9780295741826 (pbk. : alk. paper)
Subjects: LCSH: Indians of North America—Communication. | Indians of North America—Computer networks. | Broadband communication systems—United States. | Telecommunication systems—United States. | Internet—United States. | Indians of North America—Government relations—History—21st century. | Information technology—Government policy—United States. | Information technology—Social aspects—United States. | Digital divide—United States. | Sovereignty—History—21st century.
Classification: LCC E98.C73 D83 2017 | DDC 323.1197—dc23

The paper used in this publication is acid-free and meets the minimum requirements of American National Standard for Information Sciences—Permanence of Paper for Printed Library Materials, ANSI Z39.48-1984. ∞

To tribal youth:
Keep learning, and practicing listening to the mountains, the earth, the animals, the stars, as much as you read books and play videogames. Care for your families, hope, create, and embrace joy.

CONTENTS

PREFACE

Lios enchim aniavu. Inepo Marisa Elena Duartetea. In hapchi Marco Antonio Duartetea into in ae Angelita Molina Duartetea. In wai Carlos Antonio Duartetea into in wai Micaela Calista Duartetea into in wai Alejandro Antonio Duartea. In havoi Agustin Ruiz Duarteteak into in haaka Margarita Cervantes Duarteteak. In apa Juan Molinateak, yo'owe yo'owe, into in asu Maria Amacio Molina Tosaim Kavaikame. I write for my relatives, and I write for my friends, who work hard to uphold their tribal ways of life.

The word is bound to the breath, and the breath is bound to the spirit. The word is a loose bead running on a cord connecting the breath and the heart and the mind. The mind is filled with ideas, and these ideas are like stones. The stones are the children of the earth's fine inner workings, upheaved from mountains and polished smooth by rivers, oceans, and winds. Every stone belongs somewhere. Every stone comes from somewhere. Eager to please one another, human beings rush about filling their heads with ideas the way children fill a basket with stones when they go scrambling about the desert or the rocky beach.

At times I would take breaks from thinking about the impacts of the Internet in Indian Country and walk to a section of Cabrillo Beach, a shoreline in Tongva territory, off the southern coast of Los Angeles, and listen to the ocean tumbling rocks against the shore. At times I visited Lake Washington and beaches on the shores of the Puget Sound—the Salish Sea—under the gaze of Mount Rainier. For one autumn season I meditated every morning in a friend's yard filled with stones and watched the light change against the Tucson Mountains and Mount Lemmon, ranges filled with stories of pilgrimages, miracles, spirits, and survival.

Young children throw rocks at one another out of curiosity and spite. Adults throw ideas at one another out of curiosity, and sometimes also out of spite. We can forgive a child throwing a stone. It is more difficult to forgive an adult for hurling a monstrous idea at another human being. When teaching students about racism and colonialism, I remind them, "You are educated human beings. Remember that your job is to promote knowledge and wisdom, and not ignorance. Even top professors are capable of fomenting ignorance."

I based this work on the following risky ideas:

1. Human beings are also herd animals. They are capable of organizing beyond the level of the individual. They orchestrate activity at the level of a community, and articulate their identities based on geopolitical locations and status. En masse, they become swept up into communal systems of belief.

2. Human beings are inherently creative. They create systems and structures in this world through the use of tools. The physical manifestation of these systems and structures reflect human beliefs over time.

3. The present-day use of the word *technology* is laden with present-day beliefs about progress, scientific and ethical advance through computing, and the superhuman conquest of time, space, history, and environment. There is a belief that being able to speak in code—that is, programming code—parallels the decoding of the human genome and the dark matter of the multiverse, and that somehow this process of coding and decoding is meaningful for all mankind. These beliefs derive from a Western European Enlightenment history of ideas. Like a magic bullet, the word *information* can at once comprise programming code, genetic code, and the nearly immeasurable mass that one nanoparticle passes off to another when they collide in the vacuum between all other known and measurable subatomic particles.

4. The large-scale forces of Western European modernities have resulted in the creation of a global class of humans referred to as "natives," "aboriginals," or "indigenous" persons. Across modern nation-states, that nomenclature refers to a historiographical moment, when nation-state authorities were charged with classifying all resident human beings as subjects or nonsubjects, citizens or noncitizens, slaves or workers, and so on. The words *Native American, Aboriginal*, and *Indigenous* are embedded with a tension of belonging and yet not belonging to the modern nation-state. For an American Indian, it is to be called by all non-Natives an alien within one's own territory, in the shadows of one's own grandmother mountains.

5. Various fields of science are at present dominated by those who believe that techno-scientific advance must come from a Western European his-

> tory of ideas and not from, for example, Tsalagi histories of ideas, Hiaki histories of ideas, A:Shiwi histories of ideas, Anishinaabe histories of ideas, Chamoru histories of ideas, and the like. Only recently have a few scientists working in their universities come to agree that Native ways of knowing compose a source of scientific understanding. "Native ways of knowing," "Indigenous knowledge," "Native systems of knowledge"—all these phrases refer to a complexity of understanding about the nature of the human universe. As scientists—and especially as information scientists—we are only at the beginning of our understanding.

I'm Yaqui. I'm a woman writing in the sciences. I often write and research far from my home, the Sonoran desert landscape around Tucson and the northern Chihuahuan desert in southern New Mexico, and I often write through fields of science that, thus far, are inadequate in terms of language and theory for scoping the lived realities of present-day Native and Indigenous peoples. If the word is a loose bead on a cord connected to the breath, the heart, and the mind, and I am trying, from my lived experience and ways of knowing, to share that word (or words) with another human being who does not share the same ethical orientation (heart) or ways of problem-solving (mind), then what can be the significance of the word I seek to share?

The risks I have taken as a thinker are lesser than the risks I take as a writer, assembling these ideas like rocks in a basket, which I now present to you in this form, as a manuscript. This is the nature of writing. Once a story is loosed into the world, it no longer belongs to the writer. It belongs to those who hear it, and especially to those who retell it. At a certain point I can no longer insist on what is right and wrong about an idea that I have written. I can only say, "I thought quite a lot about selecting this particular idea and explaining it in this particular way." The rocks get taken from the basket, broken into smaller pieces, polished, or transferred into the baskets of others.

But what about the basket? That is the real contribution here. I am weaving a container for others to reuse. What might the Native and Indigenous peoples of the world have to say about their experiences with information and communications technologies? What might our experiences as Indigenous peoples teach us about the ways we conceptualize this ineffable, somewhat immeasurable phenomenon we pursue, which we are calling "technology"? How does it relate to our dedication—*itom herensia, itom lu'uturia*—to uphold our tribal ways of life?

I pray for the words to have meaning, and for the writing to be clear and inspiring. As the methods are true, so is the writing here.

Lios enchim hiokoe utessiavu.

Marisa Elena Duarte

ACKNOWLEDGMENTS

MANY INDIVIDUALS AND INSTITUTIONS SUPPORTED ME THROUGH the thinking, research, and writing that shapes this work. I would like to acknowledge Jose Antonio Lucero and Bret Gustafson, who, through their work as directors of the Global Indigenous Politics group with the Social Science Research Council, helped me think more deeply about early assumptions of my work. Compañeros Simón Trujillo, Cuauhtémoc T. Mexica, Shawn Walker, Jeff Hemsley, Robert Mason, and members of the SOMELAB at the University of Washington also helped me conceptualize ideas around networks and organized human action, in digital environments, in the borderlands, and in the space of civic engagement. The many scholarly conversations I enjoyed within the Indigenous Encounters forum and the Native Organization of Indigenous Scholars at the University of Washington also helped me focus on the particularities and beauty of politics and creative work in Indian Country. Of course, many thanks to the members of the Indigenous Information Research Group at the Information School at the University of Washington: Cheryl A. Metoyer, Miranda Hayes Belarde-Lewis, Sheryl Agogo Gutierrez Day, Allison B. Krebs, Juan Carlos Chavez, Sandy Littletree, and Ross Braine. We work together with one heart. My school is the iSchool. Great respect and honor to Cheryl Metoyer, Maria Elena Garcia, Raya Fidel, and David Levy for helping me prepare this work in a rigorous and readable fashion and for pushing me to think at new heights. Many thanks to Ranjit Arab, who helped edit many drafts. My family, I love you and write for us to overcome violence and tragedy, for all of us to flourish together in a joyful way, in this world and in those beyond.

This project was funded in part through the Institute of Museum and Library Services Washington Doctoral Initiative, the Social Science Research Council Dissertation Proposal Development Fellowship, the Mellon-Sawyer Predoctoral Fellowship through the Simpson Center for the Humanities at the University of Washington, the Chancellor's Postdoctoral Fellowship in American Indian Studies at the University of Illinois, Urbana-Champaign, and through higher education funding through the Pascua Yaqui Tribe.

NETWORK SOVEREIGNTY

Introduction

All night long in Room 1212 they had discussed a network of tribal coalitions dedicated to the retaking of ancestral lands by indigenous people.

—Leslie Marmon Silko, *Almanac of the Dead*

ON DECEMBER 22, 2012, NATIVE PEOPLES ALL OVER THE UNITED STATES and Canada were organizing flash mobs to protest Canadian prime minister Stephen Harper's plans to break treaty obligations to tribes in order to make way for the construction of a transborder oil pipeline and tar sands extraction. Earlier that month, to protest the unresponsiveness of Parliament regarding First Nations rights, Chief Theresa Spence of the Attawapiskat commenced a six-week hunger strike. Native peoples representing many tribes drumming under the banner of Idle No More protested in malls, parks, across major highways, before embassies, and on college campuses throughout Saskatoon, Saskatchewan; Toronto; Albuquerque, New Mexico; Tucson, Arizona; Minneapolis, Minnesota; Los Angeles; Seattle; and Vancouver.[1]

In Mexico, on December 21, the Indigenous peoples' collective Ejército Zapatista de Liberación Nacional marched en masse through five cities in Chiapas, protesting the unjust and immoral capitalist economic development policies and drug cartel violence promulgated through the administration of former Mexican president Felipe Calderon and current president Enrique Peña-Nieto.[2] People of my own tribe, the Yaqui tribe, in Sonora, Mexico, blockaded shipping and transportation between the cities of Obregón and Guaymas.[3] Two years before, a young man from the tribe used his smartphone to record and post a video of Mexican state police beating up tribal people for hauling water from a river dammed to divert the flow of water that runs through the sacred homelands. The state government agreed that the dam had been built without appropriate tribal consultation and, in return, offered to pay for university scholarships for all tribal youth. Record numbers of Yaqui youth applied and got into school. The state government reneged and refused to pay the tuition. Independent journalists posted photos online of parked semis blocking all interstate traffic passing through Vicam Switch, a predominantly Yaqui community on Interstate Highway 15 in Sonora, a primary north-south route in the country.

No more than an obscure myth for most Americans, December 22, 2012, marked the end of a five-hundred-year cycle according to the Mayan Daykeepers. But for the Indigenous peoples of the Americas, this date predicated a beginning, an opening up. Moreover, as Laguna writer Leslie Marmon Silko indicates in her 1994 novel *Almanac of the Dead*, Indigenous peoples of the Americas had been preparing for this calendrical shift for centuries.

To many Indigenous peoples of the borderlands, *Almanac of the Dead* is a guide, a manual for understanding how those of us separated by the legacies of colonialism would experience life in the interstices of cities, states, and nations; mass media and voicelessness; the regulated marketplace of ideas and the black market; guerrilla warfare and political persuasion. In the book, a group of spiritually minded Indigenous individuals congregate in a seedy hotel room in Tucson to lay out on the table, beside cigarettes and beer, their observations, insights, and visions about tribal peoples all over the Americas, including those who had settled there from Africa, working incrementally to reclaim spiritual and political relationships with their homelands, and how, at a certain point, the accumulation of this activity would visibly manifest. In short, all night long, in room 1212, they talked story.

Through the course of my education, both spiritually and politically as a Yaqui woman, *yoem hamut*, and a professional librarian and information scientist, I came to recognize the value of stories as the currency binding multiple parallel and sometimes incommensurable worldviews, including ontologically distinct Native and Indigenous worldviews. Already attuned to the deeply Indigenous concepts of relationality, interconnectedness, and emergence, as a scientist I picked up the poststructural study of networks—social (actor) networks, social media networks, sociotechnical networks—as one point through which Native and Indigenous scientific principles could inform and be informed by Western methodologies.[4] Moreover, I studied networks and the technical systems through which they physically manifest, with an eye toward Silko's prophetic reading: What are the technologies that will allow us, as Indigenous peoples, to reclaim our lands and ways of being—spiritually, socially, spatially, ecologically, and politically?

I had read the works of Western scientific theorists who predicted the identity-based mobilization of Indigenous peoples, enabled through the availability of social media. Throughout my entire upbringing, I had been privy to kitchen table conversations, conversations by fires late at night in deserts, at powwows, in aunts' living rooms, in the back of trucks, about the ways that we, as the grandchildren of elders, spiritual leaders, and military generals—survivors—would have to re-order our ways of being to usher in new worlds of possibility for Indigenous peoples. On December 22, I set aside my stack of scientific articles on technical wireless networks and traced the flurry of stories about Idle No More, the Ejército Zapatista de Liberación Nacional protest, and the Yaqui blockade of Guaymas via Facebook. These parallel Indigenous protests were not covered by CNN, NPR, or the BBC. For three years, my colleagues and I at the University of Washington Information School had been looking for contemporary examples of Indigenous peoples harnessing social media toward political mobilization. We all had examples of small-scale focused mobilization, but Idle No More's political mobilization was the first of its kind that was transnational, fast-paced, self-organizing, emergent, dynamic, and showing clear signs of generating public responsiveness. Many times before the hashtag #idlenomore of December 2012 appeared, I had imagined how Indigenous peoples might organize across national boundaries using social media, but I had never seen a transnational Indigenous political movement emerge so quickly through social media networks. From a scientific perspective, this occurrence means that by December 2012, Indigenous people throughout Aotearoa (New Zealand), Australia, Canada, and the United States—the English-speaking Indigenous world—had established multiple trustworthy and reliable mobile digital social networks across various social media platforms and devices. It meant that an aspect of Indigeneity, as a paradigm of social and political protest, had become digitized, infrastructurally through broadband Internet, personally through consumer mobile devices, socially through social media adoption, and discursively through flash mobs, hashtags, and memes.[5]

Native and Indigenous scholars have argued, mostly from a first-world English-speaking (United States and Canada) context, for Native peoples to frame the contemporary relationship between recognized tribes and the nation-state as one based on the need for Native peoples to leverage self-determination toward building a just world for themselves with regard to, and in spite of, ongoing colonization. Policies of sovereignty and self-determination are to be understood as stepping-stones toward a more flexible, morally Indigenous vision of governance.[6] At present, and in part due to the way that information and communication technologies (ICTs) have shaped US domestic and global hegemony, the leaders of Native nations must understand how information flows, the disciplining and transfer of knowledge, and technological innovation and surveillance function within the multivalent power dynamics of contemporary colonial arrangements.[7] More fundamentally, this means understanding when, where, and how autonomous Indigenous peoples can leverage information flows across ICTs for the purpose of meeting social and political goals, in spite of the forces of colonization.

As my late friend and colleague Allison B. Krebs (Anishinaabe) taught me, this is a continuation of what Vine Deloria, Jr. (Dakota), asserted during a 1978 White House presentation on library services in tribal lands. It is our right as Indigenous peoples to know the origins of our current status as colonized peoples.[8] It is our right to know this so that we can speak back to unjust governmental power. It is our right to mobilize, enact, and determine our own trajectories as Native and Indigenous peoples. While the protests of December 2012 represent a particularly striking mode of political mobilizing and government interactions, Native and Indigenous peoples have endured centuries of colonization in part because of daily ordinary habits of sharing information and ways of knowing with one another, workmates, allies, and friends. Uses of social media and the undergirding systems of devices are becoming, in Native and Indigenous contexts, common modes of sharing information and knowledge critical to Indigenous self-determination.

In the United States, the tribal command of broadband infrastructures and services represents one way that Native peoples leverage large-scale ICTs toward accomplishing distinctly Native governance goals. While these goals are particular, and depend on each tribe's ways of approaching its mode of self-government, because of the future US reliance on pervasive high-speed Internet as a means of interacting with citizens and administration, tribal leaders will want to make sure that, at minimum, tribal administration buildings, schools, and libraries have access to robust and affordable broadband Internet devices and services, including wireless capabilities. As governments, tribes possess the means for acquiring the infrastructure and services that make social media and mobile devices work from deep within Indian Country.

I write this book to (1) weave Native and Indigenous thought more firmly and productively into the broad fields of science, technology, and society studies; (2) introduce Native and Indigenous thinkers to the language of information science and sociotechnical systems; and (3) share what I have learned thus far about the uses and implications of broadband Internet in Indian Country with colleagues in the sciences, students, educators, policy makers, tribal leaders, and the general public. We have work to do, getting our people connected. We have work to do, sharing with one another the millions of stories about our strategies for effective negotiation, our failures, our visions for bringing about a healthier world for our children and the grandchildren they will one day bring into this world. ICTs are an important medium for this intergenerational and intertribal transmission of knowledge.

This book is a scientific narrative, following the arc of a scientist walking on the path of discovery with the rigor of specific methodological interventions. I am the scientist and the writer, and I am also Indigenous, specifically, a Yaqui woman with an intellectual lineage born out of the particular exigencies in the US-Mexico borderlands. I grew up hearing tribal stories and border stories. I grew up respecting the value of tribal peoples' knowledge and also the life experiences of self-made individuals, especially those who reside in the liminal and jurisdictional interstices between countries, institutions, and cities, not to mention languages and ethnic identities. I grew up attuned to the finer meaning of Native storywork, understanding how elders and other wise and thoughtful types transform the present web of meaning and therein create new possibilities for ways of being, acting, and thinking when they talk story. At present, it is still common in academia for practicing scientists to attempt to delegitimize the intellectual and scientific contributions of Native and Indigenous thinkers, as well as the participation of women. Many scientists are unfortunately not responsive to the manifestations of storywork, the interventions of feminist scientific approaches, or the intellectual claims of Indigenous research. Many scientists are financially and intellectually invested in discrediting or marginalizing the contributions of Native and Indigenous thinkers. As Indigenous thinkers are well aware, a mechanism of colonization is the subjugation of Indigenous knowledge. Through centering Native experiences and weaving together Indigenous and information scientific methodologies, this book challenges the shadow of epistemic injustice.[9]

However, this work is not intended to serve as an exhaustive Indigenous critique of techno-science; rather, it is intended to open up new ways of thinking about digital technologies, and specifically of high-speed Internet, in Indian Country. Sociotechnical conceptualizations are a relatively new analytic lens in the general field of Native and Indigenous studies. Indigenous approaches to studies of the Internet and ICTs are also new. Thus, this examination focuses on the intersection of a few disciplines, epistemological landscapes, and knowledge domains. Each chapter builds on ideas framed in preceding chapters, so that, by the end of the book, the general reader in Indigenous studies has increased knowledge about information scientific approaches and the general Internet studies researcher has increased knowledge about Indigenous experiences and approaches. Together we will have an increased understanding of the place-based nature of the Internet and how jurisdictional borders shape access to this integral communications infrastructure. This work is organized not according to the chronology of historiography but according to the way findings revealed themselves to me, the Indigenous scientist, as I began methodically asking the question "What is the relationship between Indigenous peoples and ICTs?"

This book is organized into eight chapters. Chapter 1 utilizes the example of Idle No More to introduce a basic way of understanding flows of information within the relationship between the technical and the social and how these flows manifest in Indigenous contexts. Chapter 2 explains how the scientific inability to perceive the rhythms of colonialism has led to the current need for a way of understanding the impacts of ICTs in Indian Country beyond the depiction of Native peoples as have-nots in a technically advancing postcolonial global order. Chapter 3 builds on the lessons learned in the first two chapters to explain the close tie between access to technical systems and contemporary exercises of tribal sovereignty, giving examples from five cases of Native ICTs projects. Chapter 4 looks at the inner workings of four sociotechnical systems in Indian Country and one intertribal multi-organizational forum: the tribally owned broadband Internet networks operating out of the Southern California Tribal Chairmen's Association, Coeur d'Alene, Cheyenne River Sioux, and Navajo Nation and the Tribal Telecom and Technology Summit. Chapter 5 utilizes these cases to identify the practical needs of tribes who command their own Internet infrastructures. Chapter 6 theorizes the sociotechnical dimension of future exercises of tribal sovereignty and what this means in terms of how we conceptualize the technological fabric woven throughout Native homelands. Finally, Chapter 7 approaches colonial assumptions about social studies of technology and lays out an approach that we, as scientists and Indigenous thinkers, can take in decolonizing the disciplinary practices and discourses surrounding technology studies.

CHAPTER 1

Network Thinking

Information Flows Like Water

Indigenous revival cannot be understood without reference
to the technology, power, and legitimacy of states.

—Ronald Niezen, *The Origins of Indigenism*

HOW DOES THE CONCEPT OF TECHNOLOGY RELATE TO THE CONCEPT of Indigeneity? How are the technical devices that shape contemporary day-to-day life woven into those moments that define what it means to be Indigenous? The sleek look and discreet design of many contemporary digital devices—mobile phones, laptops, tablets—invite us to imagine these objects as neutral and futuristic, devoid of historical legacies. Ironically, tribal peoples are also imagined as beings without histories: prehistoric, precolonial, and pretechnological subjects of a techno-scientific American empire. How do these parallel imaginaries weave together? How does thinking in terms of networks and relationships help us understand the way the divide between the technical and the social manifests in Indigenous contexts? Understanding the concepts of technology, Indigeneity, and networks requires an understanding of the functions that communications technology and Native peoples—Indians—completed in the formation of the modern technically advancing nation-state.

We can consider the lineage of the wireless mobile phone, before the landline telephone, before wireless telegraphy, when the railroad barons were competing in the race to build a transcontinental railroad. In the United States and parts of Canada, the late nineteenth century spelled the beginning of an increasingly industrial era of modernity, as well as a century of campaigns against Indigenous peoples. Entrepreneurs inspired by the presentations of scientists and ethnographers at the world's fairs seized on global visions of industrial, economic, and political largesse. Dreams of transcontinental transportation, communication, and shipping seemed very possible with the inventions of the steam engine and telegraphy, as well as Frederick Jackson Turner's burgeoning vision of Manifest Destiny.[1] For the landed entrepreneur, it seemed as though land and labor were there for the taking. As the railroad barons pushed westward, the US government sent agents and military backup to push tribal peoples of the western prairies, deserts, and plains from their homelands, freeing property for American settlement. By the mid-1860s, communications entrepreneurs had set up wire-line telegraph posts at many military base camps and at established train stations. Telegraphy was an important medium for transmitting messages about Indian mobilization. In the summer of 1865, a young US Army corporal assigned to guard the telegraph wire in Dakota Territory was writing a letter to his sister about Indians cutting the wires when he received a dispatch that Indians had killed the telegraph operator at nearby Sweetwater and ten military men were sent to repair the damaged line.[2] This was only a few years after Little Crow's War.[3] Stories of Indian raids were fresh in the settler imaginary, particularly in areas where the only way to achieve speedy delivery across vast western expanses had been the recently defunct Pony Express.

A decade later, and thousands of miles away, during the winter of 1874, a teenage Nikola Tesla escaped Austrian military conscription by hiding in the mountains of Croatia, hunting, hiking, and imagining the conditions under which he might be able to shoot mail through a pressurized tube under the Atlantic Ocean. Later, during the Spanish-American War (1898) and the US acquisition of a divided Samoa (1899), Tesla would set up and work in a lab in Colorado Springs, experimenting with wireless communications and envisioning invisible systems of currents transmitting whole sentences around the globe.[4] Yet, in 1901, Guglielmo Marconi, an Italian engineer, beat him to the punch, transmitting the first transatlantic radio signal from Saint John's, Newfoundland, to Poldhu, Cornwall. The race to wireless communications was in full swing, enchanting business investors and inventors alike. For Marconi, Tesla, and many others, the dream of wireless communications included hopes for an invention that would change the very nature of international politics, commerce, and human interaction with the natural world.

Meanwhile, south of Colorado Springs, the US cavalry chased Apaches along the recently settled US-Mexico border. While the world was opening up for many American immigrants, the mechanisms of national expansion were cutting short ancient proven ways of life for Native peoples in the Americas.

The general rule was that for Americans to prosper, Indians had to die. In the summer of 1905, Geronimo, semi-detained by the US Army, hopped into an automobile at the Miller Brothers' 101 Ranch Show and chased and shot a buffalo, both shocking and entertaining nouveau Americans.[5] Forty years later, a contingent of Navajo (Diné) soldiers would innovate a Diné-inspired cipher for transmitting military surveillance intelligence across radio waves in the World War II Pacific theater. Phil Deloria (Dakota) described these Indigenous encounters with technology in the American settler imaginary as "unexpected." During World War II, and for years afterward, the general US approach to tribes was termination of federal assistance to Indians, even as the United States prospered on multiple illegal claims to sovereign Indian land.[6] The practice of subjugating the land—and the Indigenous peoples of those lands—to serve the pursuit of wealth by industrialization bloomed fully throughout the Americas. In the modern settler imaginary, any Native or Indigenous use of modern technologies was unexpected precisely because Native and Indigenous peoples themselves were unexpected in the subjugated, mediated landscape. They were expected to have faded away along with shrinking herds of buffalo, made illegible by their supposed illiteracy and perceived inability to adapt to a modern industrial, and eventually techno-scientific, path to progress.

Understanding technology—and, in particular, digital technology—requires understanding the conditions under which innovation occurs. At this moment in the tapestry of world histories, digital technology is fascinating precisely because, in the technically advanced places, there remains a memory as well as some judgment of what everyday life was like before the precision, efficacy, immediacy, and interoperability of ubiquitous computing. Read against the century of US anti-Indian campaigns and imperial expansion, narratives of technological advancement function to satisfy societal desires for Enlightenment-era values of progress and scientific evolution in spite of the colonial fabric of Indian eradication.[7]

Indeed, Enlightenment-era values very much shape the concept of information itself. In 1432, the term was used to indicate matters of surveillance and accusation, and the subsequent need for adjudication, by English authorities.[8] It was a term dependent on the machinations of power politics, useful only—as all "information" is—when pieced into a greater strategy, formula, or design. With regard to the politics of sovereign authority, it was the nascent notions of empire, conquest, and saltwater colonialism that shaded the Enlightenment project as much more than state making but as part of a greater strategy of economic war making through the occupation, removal, and settlement of

Indigenous bodies, lands, and waters. The military, ethnographic, cartographic, and economic visions of the known world—even the notion that it might be possible for noble elites to know the shape and characteristics of places all over the world—depend on stores of transferrable information: maps, compendiums, libraries, and, now, databases, sites, and search algorithms. Through witnessing the rise of European fascism and alienation shaped by industrialization, French Christian anarchist Jacques Ellul critiqued the human capacity to reduce, particularize, and technologize all aspects of the lifeworld—to rely on manufactured information—until the technician could no longer separate himself from the machinery routinizing the day's labor.[9] During the Cold War, the concept of information played an important role in the development of international intelligence and security work, as well as in the innovation of the bit—the binary digit—that underlies modern computing and communications theories. Indigenous thinkers should not imagine that notions of binary mathematics, categorization, classification, accounting, cartography, technique, and literacy were unknown in places such as the pre-Columbian Americas. There were the codices of Tenochtitlán, pictographic and woven systems for inscribing and calculating trade histories and outcomes, Incan quipu, bas-relief, astronomy, and, of course, the Mayan calendar and the continuing work of the Daykeepers. Information now signifies any bits of data that can be removed from the whole, broken into pieces, refined, put back together, parsed, communicated or transmitted, and reformed for the purpose of generating new products bearing meaning.[10] It is, in itself, a practice and ideology accumulated out of the recurring need to organize strategies for knowing across competing global epistemologies.[11]

In the present communications era, in which US Army corporals guard telegraph wires far less than techno-entrepreneurs lobby around Net Neutrality, privacy acts, and nondisclosure agreements, there is a language for orienting oneself in the world of information.[12] Data, parsed through systems of devices, become information. Information accumulates in myriad ways to form the basis of knowledge. Humans, aided by institutions that sanction and distinguish information from knowledge, canonize, so to speak, what is knowledge. Individuals working through libraries, churches, the free press and the state-run press, schools, universities, hospitals, the state, research labs, and think tanks separate the wheat from the chaff, and the products of their parsing, synthesis, and reformulation emerge most often in the tangible and visible form of documents. The data that describe these documents—books, papers, and, nowadays, sound files, websites, and databases—are known as metadata.

If we were to reenvision the world as a series of ones and zeros, machine-readable and system-compatible, we might tend toward a view of the universe in which changes occur not as an outcome of social forces or great heroes and institutions but rather due to the algorithmic accumulation of information through interlaced networks of humans and devices. Data gain in value insofar as they are relevant to a whole. Information gains in value based on its increasing circulation within greater circuits of meaning. Knowledge is negotiated, sanctioned, agreed upon by authorities, guarded by authorities, and priced by authorities. Did Marconi really outwit Tesla, or was Marconi affiliated within networks of individuals, institutions, and devices that created the conditions for speedier project completion? (After his death, investigators found that many of Marconi's patents were dependent on Tesla's original inventions.) Did the Code Talkers really develop a completely new cipher, or did they simply bring their knowledge of their homelands to bear, bringing a fresh perspective based in ancient ways of knowing into a hierarchical military unit that had not before imagined—that had not expected—that critical information about enemy movements could be inscribed in such a way?

When most people speak about technology, it is most often in reference to a specific device: an object, like an Xbox, iPhone, or tablet, for example. Many times individuals speak about technology as if it were an object that could be removed from the immediate environment with no effect. Consider the mother who does not allow her teenage daughter to own a smartphone. Yet when this same teenager goes to her high school lunchroom in Palo Alto, California, she finds herself surrounded by friends and perceived enemies, who all own smartphones attached to Facebook, Twitter, and Snapchat accounts. Suddenly, the world of teenage relationship building and identity formation is mediated, even in unmediated places, such as beneath the shade tree just outside the lunchroom doors.[13] Earlier information scientists imagined terms and languages for describing this digital overlay in everyday life, in which the individual's distance from a device does not compel an existence free of technological conditions and the social power abetted through the continual circulation of digital information.[14] A technical device and the information flows that abet its use are predicated on layers of human relationships and are imbued with power.

While it is possible to frame technology as a societal condition, an environment of industry and innovation, a social scientific mechanism, or even, popularly, a kind of machine, as an information scientist I think of technology more often as a system of devices through which information—legible, respectively, to humans and machines—circulates to give depth or add meaning to orchestrated human goals. Designers create devices based on previous innovations, and for specific work purposes. These work goals are laden with human values—values about human interactions, the environment, approaches to work, and so on—and these in turn shape the design and uses of devices. The interfaces—that is, the surfaces of the devices that make digital information legible to humans—function like sieves, filtering noise from experience.

Through the design of interfaces, there is something to be learned about human interaction with information, across affective, cognitive, behavioral, and sensory domains. Smartphone designers create interesting new interfaces and test human interaction with these interfaces across groups of users. The language is at once both hopeful and predictive: the human who approaches this device—whether or not he or she has ever seen it before—is already socially conditioned to tune in to the nature of its purpose, functionality, and utility and so is termed a "user." Indeed, the user is already a participant within a greater technological social order, in which the concept of data and information—the concept of the parsing and generalizability of meaning—exists, and in which this process of parsing and reformulation is at once both valuable and fairly free-flowing.

Indeed, it is helpful to think of information like water. At the molecular level, it is invisible and imperceptible to the human eye, yet as beings made in large part of water, we are literally swimming in dilutions of it every day, through our breath, the daily ablutions, and cups of coffee and tea. Like water, information takes the shape of its container. The job of the interface designer is to create the container in anticipation of human desires. Like water, information circulates through human-made networks oriented to fulfill communal needs. Like water, information is measured, dammed, bought and sold, regulated, and recycled. Like water, information has a value in global circuits of trade.

Yet though the idea of information flowing like water is helpful, it does not go far enough to depict the complete malleability of information, nor does it accurately reflect what Manuel Castells has articulated as the power of information flowing through networks.[15] As an Indigenous thinker, nourished from before birth by a sensitivity to the *lutu'uria*, my people's orientation to our spiritual and historical Truth, and what Subcomandante Marcos refers to as an Otherly reading of histories, I am interested in what the experiences of Native and Indigenous peoples teach us about information, technology, and the power of information flowing through sociotechnical networks.[16] More specifically, I am interested in how Native and Indigenous peoples leverage information and technology to subvert the legacies and processes of colonization as it manifests over time across communities in many forms.

Here is where the scientific understanding of networks clashes with the Indigenous experience of networks. A scientist may be able to look at a network map of, for example, the Internet service providers across the western United States in 1998 and observe features such as the asymmetry of distribution of service across regions or the density of connections in cities. A tribal person residing outside Tuba City, Arizona, in 1998 has a personal experience of the depth of that asymmetry. She races to get her asthmatic mother to the hospital in time because there is no 911 phone service out where they live, and when she tries to deal with the emergency medical team, they have no way of obtaining the necessary medical records because the Indian Health Service clinic does not have the records in digital form, ready to transmit online. She is told by the billing professional that she and her mother are "out of network." The clinical health experience becomes laden with all kinds of information asymmetries, from language barriers, to credit challenges, to misinformation, and perhaps even to racist stereotypes of the kind that occur in reservation border towns. Being able to visually map networks and flows of information—and even to strategize the construction of clinics, telecommunications infrastructure, roadways, and wide area networks on this basis—depends on an empirical experience that is distinct from that of the Indigenous person whose life and whose relatives' lives are shaped by centuries of marginalization, or, in information scientific terms, multiple kinds of network exclusion, multiple kinds of information asymmetries. As Anna Munster writes, emphasis on the "tidy" management of networks "deadens the sense of complexity" that is the life inside the network.[17] For Native and Indigenous peoples, and for Indigenous thinkers, the experiential awareness of the global social exclusion that in large part propels Indigeneity is precisely what gives them insight into the features of domination. We do not just infer the existence of exclusion; we suffer from it, and our ways of knowing are shaped by the complexity of that experience.

For example, in 1999, cohorts of psychologists and mental health workers issued a call for research supporting what they perceived as a unique array of psychosocial factors affecting Native Americans and having something to do with the more brutal experiences of reservation life: addictions; physical, verbal, and emotional abuse; high rates of depression and youth suicide; diabetes and associated heart conditions. In 2009, Dr. Maria Yellow Horse Brave Heart (Hunkpapa, Oglala Lakota) gave a lecture on the phenomenon of historical trauma and its manifestation in the physical and mental health of contemporary

Native Americans.[18] She described how a team of researchers detected a pattern in the stories that Native patients told about their family histories. The researchers found that a particular array of conditions occurred in individuals, depending on their parents', grandparents', and great-grandparents' experiences of forced relocation from their original homelands. It could be mapped through genealogy, the potential for illness predicted along that most familiar network map: a chronological family tree.

Dr. Brave Heart described with clinical accuracy the nature and manifestation of one phenomenon after another, making it clear that Indigenous people had been internalizing the very real psychosocial impact of earlier colonization policies: social policies abetting Indigenous erasure and subjugation; federal and state bans on speaking our Native language and even the Spanish language in favor of American English; relocation, refugee, and resettlement programs; the legal theft of agricultural lands, water, and even horses and cattle; military and agribusiness contamination of land and water; and, in the case of my tribe, the official twentieth-century Mexican genocidal campaign against all Yaqui Indians and their right to live in their sacred lands.

Dr. Yellow Horse Brave Heart's point cannot be understated: to work against colonialism is not simply a matter of protesting the World Trade Organization, building schools in far-off Indigenous locales, and practicing policy reform. One of the most nefarious dimensions of colonization is how it takes root in the psyche, limiting intellectual vision and poisoning the spirit with defeat by white supremacy. To work against colonization means finding these defeated parts of the self, acknowledging and identifying the nature of the struggle, and then transforming the defeat and disarray into resilience and broader right vision for the self, the family, the tribal community, and Native and Indigenous peoples.

Thus when Native and Indigenous thinkers position themselves to examine flows of information, networks, and the social impacts of devices, we do so with an eye toward social transformation and change. While an important initial stage of research in this regard is mapping information flows and articulating the phenomena occurring around an ecology of devices, a key subsequent stage is identifying what this means for Indigenous thought and praxis. How do we change the conditions of our exigency?

The colonial experiences of contemporary Native peoples in the United States are similar to, but not a replication of, the experiences of contemporary First Nations peoples in Canada, and neither is identical to the colonial experiences of contemporary Indigenous peoples in Mexico. Indeed, the experiences of Native peoples in the United States vary by region and era, by tribe, and by the unique philosophies of the more than six hundred Native peoples whose original homelands are within the current sovereign US jurisdiction. This means that, theoretically, there should be more than six hundred unique and related manifestations and experiences of colonialism, based on the range of colonization policies effected over generations of Native and Indigenous peoples who were forcibly removed during US expansion and who continue to grapple with affiliated policies through the present.

Furthermore, there are a myriad of distinctly non-European Indigenous philosophies shaping contemporary values, attitudes, and behaviors—and guiding the shape of tribal institutions—in the United States alone. Though a familiar concept to students and educators in American Indian Studies, and an ancient and fundamental concept for tribal elders who speak and think in the original languages, this concept is dazzling and complex for nationalist thinkers who identify Others on the basis of country of origin, then ethnic group, then state or province. It is like discovering a whole Other order of governance (and chaos) beneath the federal government. I am often asked, if there is such difference and uniqueness among Indigenous peoples within the United States and throughout Canada and Mexico, then what is the basis for the US ethnic minority subgroup known (as of late) as "Native Americans"? What is the basis for the term *Indigenous peoples*?

The short answer is that Native and Indigenous peoples bond through a shared experience of overlapping waves of colonialism. Colonialism is the social condition—the milieu—that abets the enactment of colonization policies. The contemporary goals of colonization are fundamentally tied to the goals of an elite class of citizen-subjects who desire to remove the original peoples of a land from that land in pursuit of the greater goals of settlement and nationalist modernization or development. The process of removing, relocating, and eradicating the original peoples—the Native and Indigenous peoples—and then subjugating the proof of Native claims results in a regenerative violence, productive for elite and landed middle-class settlement efforts.[19] With Native and Indigenous modes of social governance in disarray, and the people dying, the elite can develop the lands and waters in ways that serve the erection of institutions and practices that support elite class goals.[20] When Ron Niezen writes, "Indigenous revival cannot be understood without reference to the technology, power, and legitimacy of states,"[21] we can conceptualize how colonial authorities work through systems of analog and digital devices, rules of law and customary practices, state institutions, and the flows of information that bind these together. In turn, Indigenous peoples resist and subvert colonizing systems, rules and practices, and state power through social and political engagement and mobilization. Indigenous peoples conscript digital devices and systems to do this work, and it is not necessarily in conceptual contradiction to Indigenous philosophies, spiritualities, and everyday practices. It is not an unexpected practice; it is becoming a technique of Indigenous praxis.

Unlike most information scientists, Indigenous scientists are interested in how information and technology shape the colonial milieu. We pay attention to the ways knowledge is articulated in order to acknowledge, reveal, denigrate, or elide Indigenous experiences and make mental note of data sets that account for or do not account for Native peoples and how the availability of data about Native and Indigenous peoples shape decision making in and around Native communities. We think about the devices, systems, and networks through which information flows in and around Native communities. Relating systems of devices and information flows to colonialism requires integrating the material and tangible aspects of information systems with sociological explanations of the mechanisms of colonialism. It requires reading Indigenous theories of colonialism and decolonization through what Ellul would call "a technician's eyes."[22]

While there are many explanations of colonialism, few lend themselves to such subversive readings of "technicíse," or "technicization."[23] Peruvian social scientist Anibal Quijano's coloniality of power, or *la colonialidad de poder,* is one such explanation. Quijano's formulation emerged out of a decade of work with Latin American sociologists who considered the widespread effects of Spanish colonialism in different parts of Latin America over time.[24] Generally, in the context of the Spanish invasion and occupation of the Americas, colonialism manifested through four interdependent mechanisms. The first is the reclassification of the population according to a racialized, gendered, caste system that upheld the Spanish nobility as elite authorities and denigrated slaves and Indigenous people (*indios*) as the least powerful class. This classification contributed to the second mechanism: the articulation of institutions such as churches, schools and colleges, government offices, medical offices, banks, *haciendas* and *ejidos*, and marketplaces, which reified the economic and social power of the Spanish elite while reducing the lower castes as labor at best. The organization of these institutions depended on the third mechanism, spatial redistribution, in which colonial authorities ravaged and destroyed Indigenous places of worship, fields, buildings, schools, markets, and villages, laying claim to Indigenous places in the name of the queen, the Christian God, or the pertinent colonial authority. In turn, spatial redistribution,

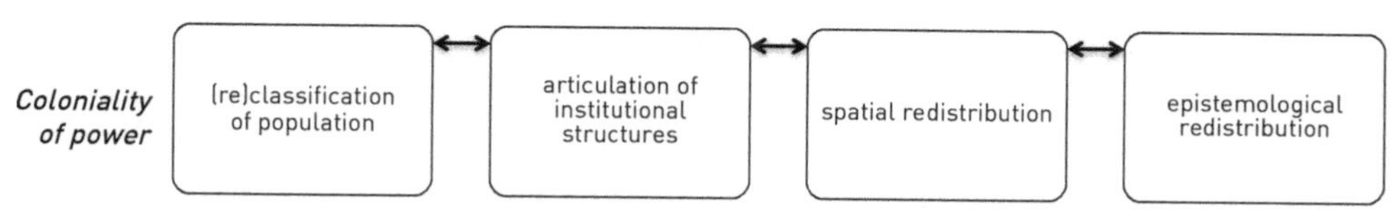

FIGURE 1.1. Four interlocking mechanisms defining a colonial field of power (Quijano, "Coloniality of Power, Eurocentrism, and Latin America"). These include the (re)classification of the population in order to sustain colonial authority, the articulation of institutions that uphold colonial work practices, redistribution of Indigenous lands and social spaces for use by the colonial elite, and redistribution of Indigenous ways of knowing, or epistemology, for the purpose of legitimating colonial knowledge production.

the articulation of colonial institutions, and the classification of the Indigenous population contributed to the fourth mechanism, epistemological redistribution. Figure 1.1 shows how these four mechanisms were dependent on one another and, together, reified the colonial field of power.

Colonial church authorities forbade and denigrated Indigenous philosophical, religious, and spiritual practices, marking these as idolatrous and blasphemous against the Roman Catholic order. Colonial schools challenged Indigenous philosophies, conscripting Indigenous scribes and translators to codify languages for the purpose of eventually using languages as tools for Christianizing and Westernizing the Indigenous intellect. Colonial authorities renamed Indigenous places with Spanish place-names, many with terms that signified places of governance and order in Spain. Spanish authorities prohibited gatherings of Indigenous leaders in public baths, public markets, and other places where provocative ideas spoken in tribal tongues might cohere and take root, threatening the colonial arrangement.

Quijano, Walter Mignolo, and many other Indigenous, European, and Latin American social theorists have applied the idea of the coloniality of power in order to understand multivalent power effects of eras of colonialism, including economic impacts, social impacts, political shifts, and mechanisms of institutional authority as these emerge over time and through particular geopolitical locales.[25] This formulation provides a frame for distinguishing multigenerational macrosociological mechanisms from particular colonizing acts. For example, the Dawes Act of 1887 is representative of the mechanism of spatial redistribution. The programs that took Native children from their families and forced them into boarding schools and missionary homes is representative of the mechanisms of population classification, articulation of church and state institutions, and epistemological redistribution. Creating maps and surveys that show no trace of Indigenous territory or terrain is representative of the mechanisms of spatial and epistemological redistribution. In sum, the effect of colonizing acts is devastating for Native and Indigenous peoples.

When we conceptualize these mechanisms as contributing features of various colonial milieus, each of which differs based on time, place, actors, and rules, we can see how the underlying mechanisms reproduce and replicate more colonizing acts, giving shape and form to what we now recognize as neocolonial strategies. Furthermore, distinguishing colonial mechanisms from colonizing acts helps us articulate the concrete, specific, material, and tangible assemblages of policies, practices, institutions, devices, actors, and information flows that compose the systems that make such acts feasible.

Colonizing systems—orchestrated units of devices, data, technicians, techniques, and rules—are the outcome and reflection of larger objectives, goals, and imaginaries pervading colonial institutions. One way of identifying colonizing systems and assemblages is to identify documents and artifacts that are evidence of the coloniality of power in motion. For example, the *casta* paintings of New Spain depict colonial classification based on blood purity, which had an effect on eighteenth-century taxation and landownership, as well as epistemological notions of self and belonging. Juan Patricio Morlete Ruiz painted many images of members of the most commonly recognized sixteen castes, although colonial authorities developed lists of more than one hundred racial and socioeconomic groups by the early nineteenth century. In *X. De español y torna atrás, tente en el aire*, the adult male figure's clothing, European features, and sword indicate his Spanish heritage and colonial authority, while the woman's clothing denotes her Moorish descent (fig. 1.2). Their child, lifted playfully between them, connotes the uncertainty of inheritance and status through the vehicle of racial admixture under a colonial caste system. Splitting tribal social orders into a colonial caste system significantly eroded Indigenous sociality, in particular since the lower castes were largely composed of Indigenous peoples.

One can research the taxation ledgers, land deeds and treaties, associated maps and surveys, and caste schedules, and the systems and devices used to create and organize this classification, and proceed to identify the parameters of a particular sociotechnical assemblage—or a set of interlocking colonizing systems and the institutions they work through—devised to maintain the Spanish colonial class hierarchy. Various kinds of information systems abet the oppression of Indigenous peoples through the colonial field of power. Figure 1.3 shows how the mechanisms of coloniality result in the design objectives of technical systems and their resulting sociotechnical assemblages.

FIGURE 1.2. Caste paintings, or *pinturas de casta*, were a genre of work in the eighteenth century in colonial New Spain. Artists depicted members of different social castes, or *castas*, as these were created through European classification methods. Castes were based on degree of Indigenous, Spanish, and African blood, in addition to maternal and paternal castes. In this painting, the European father holds a sword, indicating his noble Spanish colonial status. His wife's clothing indicates her Moorish descent. According to the caste system, their son would have had less social and political privileges than his father, but likely more than his mother. The realistic depiction of the fruit at their feet indicates the Enlightenment-era naturalist focus on classification of flora through exploration of the New World. Juan Patricio Morlete Ruiz (Mexico, 1713–1772), *X. De español y torna atrás, tente en el aire* (X. From Spaniard and *Torna Atrás, a Tente en el Aire*), 1760; oil on canvas, 39½ × 47½ in.; Gift of the 2011 Collectors Committee, M.2011.20.3, Los Angeles County Museum of Art.

Colonizing systems include systems designed to classify populations in a way that favors white supremacy and encourages Indigenous subjugation and eradication, for example, through denial of rights based on blood quantum rolls or citizenship status records. Colonizing systems also may include systems that improve the efficiency of data and information exchanges across settler institutions while intentionally circumventing Native governance and social

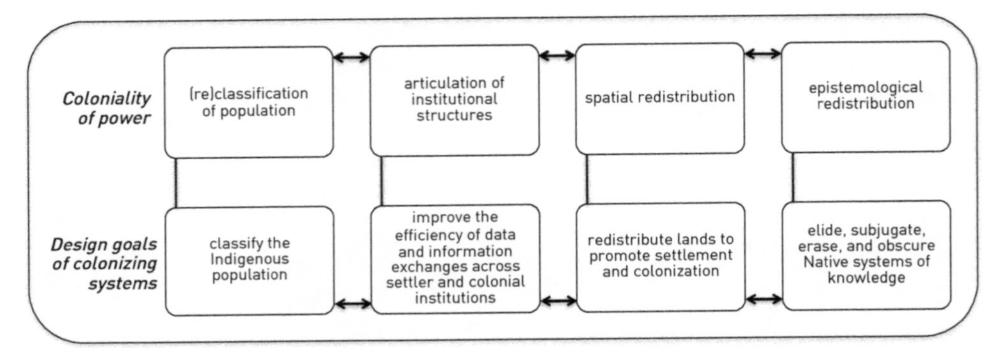

FIGURE I.3. The mechanisms enabling colonial authority require the development of information systems, most often recognizable as surveys, lists, typologies, diagrams, charts, maps, accounting ledgers, deeds and titles, and other such means of maintaining colonial bureaucratic order. These systems have design goals that uphold colonialism. Colonizing systems are designed by colonial authorities to classify the indigenous population, improve the efficiency of data and information flowing through colonial institutions, distribute lands to promote settlement and colonization, and, finally, to subjugate Native ways of knowing.

institutions. The Navajo person rushing to get her mother to the hospital because there is no 911 service on or around the reservation is one example. Eloise Cobell's remarkable accounting work revealed the inner workings of federal governmental bureaucratic systems that redistributed land from Native American and African American people to American settlers.[26] Indigenous educators consistently work against systems in public schools that prohibit, elide, or subjugate Native and Indigenous histories and philosophies. The publishing, editorial, and distribution systems that produce and promote Eurocentric American history narratives in public school textbooks are an example of a colonizing sociotechnical assemblage.

The pervasiveness of these systems can produce what feels like a total colonial experience for many Native and Indigenous individuals. From this stance, we can understand and empathize with, for example, Native individuals who refuse to allow their photograph to be taken by American anthropologists because they recognize how the camera, the photo, the anthropologist, the database, and the gallery will likely be weaponized to work against Native and Indigenous peoples in the United States while also granting the anthropologist more power within his or her predominantly white institution. The anthropologist's camera becomes a tool within a greater colonial assemblage. Native people know that many Americans are either ignorant of their role within the greater mechanisms of colonialism or, at worst, proudly profit from what they believe is their destiny and right to subjugate Native peoples. From this perspective, we can also empathize with the younger generations of Native and Indigenous peoples who connect with one another—exchange information—via social media and come up with strategies for combating what they experience as violent colonial excesses.

While a fair number of scholars have written about the varying functions and roles of social media for political mobilization, from the Howard Dean campaign to the Arab Spring and the Occupy movement, few have written about uses of social media from an Indigenous—and non-statist, non-nationalist—perspective.[27] I share with you here an Indigenous interpretation of the sociotechnical networks through which anti-colonial Idle No More activists beat their drums.

In October 2012, Canadian prime minister Harper issued federal omnibus bills C-45 and C-31, threatening the well-being of First Nations waters, landscapes, women, and children by curbing rights to sovereignty, territory, environmental protection, and treaty obligations.[28] On November 20, 2012, activists Nina Wilson, Sheelah McClean, Sylvia McAdam, and Jessica Gordon hosted a teach-in, leading to the launch of the Idle No More website and Facebook events and announcements. On December 10, Amnesty International's Human Rights Day, Chief Theresa Spence of Attawapiskat commenced a six-week hunger strike to protest deteriorating rights in her reserve. Within ten days, Native peoples and allies across North America were hosting ten-minute flash mobs—with prayers, songs, drumming, and round dances—in city malls, at parks, and in public plazas. All of this was visible in only two ways: in person and through the social media networks created by Native and Indigenous activist Facebook and Twitter users. The major news stations did not cover or connect these stories and their common impulse, perhaps in part because this common impulse has to do with the fundamental invisible work of Native and Indigenous peoples in standing idle no more in their sacred work of revitalizing relationships with homelands, waters, and lifeways by defeating colonialism.[29]

On a gray and rainy day in January 2013, around three hundred individuals, the majority of whom were Native women, gathered at the Peace Arch State Park at the US-Canada border for an Idle No More prayer rally. It was not a protest but rather a spiritual gathering at the borderland, a space where, as scholar Audra Simpson (Akwesasne) notes, the constraints of what it means to be a legitimate and recognizable citizen Indian are at once contested and reinforced.[30] Women sang and played their hand drums. There were few speeches. There was no stage or standing microphone. There was a massive round dance, and in the middle, women sang. But unlike other Native spiritual gatherings I have been a part of, this one was heavily recorded by people with smartphones, including myself, and independent journalists with lightweight cameras. The moment was a compelling intersection between systems of technical networks and various Indigenous spiritualities and intentionalities.

Near the end of the rally, the women who organized it called for Chief Shawn A-in-chut Atleo, hereditary chief of the Ahousaht First Nation, to offer closing words. Chief Atleo was not present. A young woman, a relative from those homelands, stepped up and offered a song. She was a bit reticent at first, in her black sweatshirt, holding her hand drum, telling the crowds of women, men, and journalists assembled in a ring around her about the song. She said she learned it when she was in grade school, and she never understood why they made her learn this song. It was not until later in life that she understood the meaning of the song. She drummed and sang a long and mournful song, an acknowledging song, singing, "I cry, I cry when I see where I used to walk." She turned in a circle as she sang. At one point, her voice strong, she held up her drumstick. The momentum built. The women around her holding hand drums sang and drummed with her until she closed the song. The singing offered a coming together. Her voice was like the center of the nest woven that day by the activists, tribal people, and journalists who showed up to observe, participate, and document this single event out of the myriad of moments of activism shaping the Idle No More movement.

The scientific literature lacks an explanation of the creative spiritual, emotional, and political dimensions that shape Native people's choices of select information and communication technologies for select purposes. For months, every time I watched or shared this video, I would cry, and the people I shared it with would become quite emotional. I shared it with friends and family in southern Arizona and New Mexico who do not use social media; this is how some of them learned about Idle No More. I've shared it with fellow information science researchers and Indigenous women activists. The content is like a mirror, showing us how we feel about missing a landscape so completely and having to watch it every day as it is mowed through, paved over, its waters dammed, and the life therein sickened by contaminants. It shows how as Native peoples bearing a sacred responsibility to a landscape, we feel powerless to stop this advance. It also shows the strength of Native women, who, in spite of facing radically higher levels of interpersonal and environmental violence than almost any other social group, and especially in borderlands, chose to organize through peaceful means. Idle No More flash mobs were a massive intertribal, transnational prayer rally, harnessed via discretionary use of mobile devices.

One of Idle No More's greatest strengths is, not only as a movement, but as a first visible manifestation of the sociotechnical networks created by Native and Indigenous peoples. The content is provocative, but it is the networks of devices and women using them that are powerful. This takes previous assumptions of Indigenous women as being mere labor—bodies, biopower, sources of Indigenous knowledge—in global circuits of trade and turns them upside down. It challenges the assumption that the Internet, broadly, must always be a tool of domination wielded against Indigenous ways of being. Here, Native women are commanding a section of the interface layer of global information flows so that they can set a political agenda against the unjust policies of a first-world economic and political regime. First-world technologies become, not a weapon, but a means of working toward decolonization. What I saw in this particular moment was Native women creating Indigenous cosmovisions that allow for the spiritual significance of devices. This reveals the materiality of contemporary and future Indigenous movements. Conceptualizing the ways Native and Indigenous women activists relate to one another via networks of devices helps us nuance the fabric of Indigenous activism, in which each person and each device is only a small part of complex interrelated worlds, some off-line, some online, all coherent, dynamic, and mutable in their own ways. Tracing the information that flows through these networks of individuals and the networks of devices that they build reveals the sociotechnical strata through which Native and Indigenous peoples work against forces of colonialism.

CHAPTER 2

Reframing ICTs in Indian Country

Moving beyond Cultural Difference

The issue is not about what you write with, but the hand that dreams when it writes. And that is what the pencil is afraid of, to realize that it is not necessary.

—Subcomandante Insurgente Marcos, "The Hand That Dreams When It Writes"

IN 2005, MAKERE STEWART-HARAWIRA (MAORI) PUBLISHED HER BOOK on Indigenous responses to globalization,[1] asserting that any theory that does not account for the political exigencies of Indigenous peoples may be considered incomplete. In 2010, after a few years of digging through information science, computer science, and Internet studies literature for any conceptualizations of Indigeneity—beyond the mere use of Indigenous research subjects, another brand of Othering in North American racial formulation—I realized it was time to point out and pin down the contours of this gap. What was the nature of the relationship between Native and Indigenous peoples and widespread uses of information and communication technologies?

To my colleagues and me in the University of Washington Indigenous Information Research Group, the gap in the literature was so apparent that it was visible. We could go to conferences and lectures, and read articles, and predict what misreading of Native American experiences or aspect of US colonialism the speakers or writers would promulgate. In 2010, I could speak freely and easily with my colleagues—Cheryl Metoyer (Eastern Band Cherokee), Miranda Belarde-Lewis (Zuni/Tlingit), Sheryl A. Day (Chamorro), and Allison B. Krebs (Anishinaabe)—about a range of issues that Native and Indigenous peoples were encountering with regard to media misrepresentation, lack of information critical to self-governance, and moves for autonomy borne through digital media channels. Each of these scholars had years of experience working with tribal libraries, museums, archives, Indigenous language and cultural revitalization, and systems design. They also had a strong knowledge of information institutions like libraries, museums, and archives, as well as awareness of the limitations of published academic literature with regard to Native and Indigenous ways of knowing. Thus, though the published literature was yielding a few narrative threads, it was through our talking together and thinking together that we had really begun weaving a new way of thinking about Native and Indigenous peoples' experiences with information.

Indeed, it was during a 2010 visit to my late friend and colleague Allison B. Krebs's Seattle apartment that I realized we were on the right track in our new way of thinking. A few months before, I had stumbled across a book by Mexican American philosopher Manuel de Landa in which he describes institutions as crystallizations of human ways of communicating with one another and within dynamic, ever-changing environments.[2] I most appreciated this idea for how it echoed Native concepts of creation, in which all forms that come into existence are understood as the outcomes of an endless cosmic dynamic, of which humans are a very small part. To create is to bring into being. Any object created by human hands is actually a physical manifestation of generations of conscientious human experience within a Native homeland.[3] Other Indigenous scholars who read the book have said de Landa's use of geologic time and metaphors resonated with them.

The difference between, for example, Mayan ancestors inscribing prophetic histories on a rock face and Zapatista Subcomandante Marcos issuing cyber-communiqués via airwaves is in the choice of media and the desired impact. Across centuries, however, the drive toward Mayan autonomy is the same. The philosophies are resilient, explanatory, and intact. The peoples are connected and waiting for the messages. Generations ago, Mayan ancestors learned a language, assembled a set of tools, and carved meaning into a rock face. Generations later, their granddaughters learned to program, assembled a series of laptops and radio equipment, and carved meaning into the airwaves flowing from the mountaintops of Chiapas to homes in Chicago, Mexico City, and Los Angeles.[4] The premise of Silko's prophetic narrative *Almanac of the Dead* is of a network of tribal coalitions working toward a total spiritual reclamation of the Indigenous Americas.[5] But how exactly might these intertribal networks physically manifest?

At present, there are no published theories or conceptualizations in the fields of information science or Native and Indigenous studies that center Native and

Indigenous peoples' experiences with ICTs at an epistemological or ontological level. There are descriptive studies.[6] There are narrative accounts.[7] There are approaches from the fields of communications and anthropology.[8] Yet none of these attain a level of detail that captures the richness of an Indigenous socio-technical experience. Part of this has to do with an unfortunate intellectual inheritance, the idea that Native peoples are premodern and antitechnological.[9] This colonizing logic most often emerges from works by elite nationalists of technologically advanced and rapidly industrializing countries for whom science and computing technologies have become intertwined with notions of progress.[10] It is this same logic that compels nation-state elites to relocate or eradicate Native peoples because the value of their "Indigenous knowledge" or "traditional knowledge" on the world market is greater than the value of their freedom to live in right relation within their homelands. It is this same logic that prevents scientists from understanding Native approaches to design, storytelling, medicine, and food practices as rigorous modes of communicating information and knowledge critical for human survival and resiliency across generations.[11] This blindness can be characterized as a symptom of epistemic injustice, in which the inability to see or perceive a phenomenon has less to do with honest ignorance and more to do with generations of learned habits of arrogance, laziness, and closed-mindedness.[12]

In the fall of 2009, I was very much aware that I was attempting to write about Indigenous approaches to ICTs in Seattle, one of the top tech cities of the world, ironically named after the leader Chief Sealth, whose peoples' homelands continue to be unrecognized. Though it was a temptation to presume that ubiquitous computing would encourage Indigenous peoples across national borders—Mexico, Canada, and the United States—to collaborate, I was also aware that the realities of uneven digital access and distinctive political climates and goals situated the hope for digital solidarity in the availability of affordable phone plans and fiber optics across challenging epistemic, jurisdictional, and geopolitical terrain. As users, we experience the Internet as a series of cloud-based apps, but from a systems design perspective, it is still a series of interlocking and interoperable closed systems, and though, as Indigenous peoples, we strive to think across national borders and imaginaries, the mundane acts of our lives are in many ways still circumscribed by national infrastructural capacities. Thus I realized I needed to step away from the university and open my senses to the stories of ICTs coming from within Indian Country. I wanted to understand how particular systems of devices bumped up against the constraints of tribal sovereignty and limited Internet access as these play out in specific US contexts. I needed to see the landscapes around me as an overlay of digital interactions interlacing landscapes cultivated by the hands of Native peoples working together over centuries. Rather than piecing together broken fragments, I was weaving together many narrative threads, including my own as a Yaqui information scientist working through the colonizing logics built into the research university environment.

Linda Tuhiwai Smith's handbook *Decolonizing Methodologies*, written for Indigenous researchers, seeks to heal colonial traumas in Indigenous homelands.[13] Meanwhile, I selected reframing as the guiding methodology for this work because it is a process that subverts a social problem often diagnosed as an "Indian problem" and thereby shows that it is actually an outcome of overlapping patterns of colonization.[14] With regard to the Internet in Indian Country, previous studies were diagnosing limited Internet access on American Indian reservations as an outcome of the inadequate infrastructure, remote geography, and insufficient market demand endemic to reservation life—in other words, limited Internet access on reservations was an "Indian problem." But these descriptions of Native uses of ICTs did not account for the legacies of colonialism, exigencies of tribal sovereignty, histories of self-determination, and the realities of day-to-day reservation life. In the spring of 2011, I commenced an exploratory qualitative study into Native uses of ICTs, specifically with regard to how these intersect with expressions of tribal sovereignty. The goals of the study were multiple, but at a theoretical level, I understood the need to tease out the distinction between variations in epistemic lack of awareness, distinguishing an honest lack of awareness—say, among Indigenous scholars who may be unfamiliar with web epistemologies—from the arrogance that prevents dedicated inquiry.

Within a year, the study had blossomed into an iterative qualitative study consisting of four stages and focusing on the deployment of tribal broadband Internet networks, the large-scale ICT infrastructures that enable the functioning of smaller, localized information systems and devices. The first stage consisted of reframing understandings of ICTs in Indian Country, and specifically of broadband Internet networks in Indian Country. The second stage mapped tribal strategies for acquiring broadband Internet access against the backdrop of US federal broadband deployment efforts. This was important because many Internet consumers are not fully aware of the degree to which their Internet access has depended on federal subsidy and apportioning of funds. The third stage compared four cases of self-sustaining tribal Internet service providers, each of which continues to represent a remarkable accomplishment given the near-monopolistic command that the top US providers now claim over spectrum, lobbying, and service regions. The last stage took a step back, to gain a sense of the bigger picture. How do tribal broadband networks intersect with theories of the sovereign rights of tribes and ongoing self-determination and decolonization efforts? What does the case of tribal broadband networks reveal about information scientific accounts of how power operates across ICT infrastructures? To decolonize settler-centric methodologies requires asking research questions and staging research steps, the results of which will yield results that contribute toward explaining tribal perspectives and experiences. Figure 2.1 depicts how these stages frame one another.

For three years, I reviewed the literature on ICTs in Indian Country, attended workshops, and conducted interviews and site visits with people acquiring broadband for reservation communities. I also conducted archival analysis: reviewed policy papers, broadband grant and loan applications, and infrastructural deployment plans. Building case studies out of narrative accounts of tribal Internet service provision efforts, I was compelled by visualizations of network maps, anecdotes of intertribal political organizing, southwestern Native peoples' stories of Spider Woman, and the understanding of broadband network towers emerging out of people's generations-long relationship with living landscapes. I followed the ways people used devices like smartphones and tablets and collected ephemera on Native websites, ICT businesses, and artworks. I treated the methodology of reframing as the construction of a loom holding the narrative threads in place. My writing became a design process. The ability to step back and theorize became a matter of gazing upon a fabric woven out of people's experiences written within the histories of particularly Indigenous sociotechnical landscapes.

By the second stage of the research, I had gathered sufficient evidence to recognize that the narrative threads were revealing strategies tribal leaders had developed for acquiring broadband Internet access for their reservation communities. I began identifying the problems that these strategies generated and resolved, as well as the social and political impacts of these strategies. In sum, I understood that these strategies could help us foresee, as information scientists and as scholars of tribal sovereignty, the implications of deploying a major US ICT infrastructure across sovereign tribal lands.

Ultimately, reframing is a powerful methodology because it accounts for Native peoples pushing beyond the colonial boundaries that have curbed their ability to share information and knowledge through the media of ICTs. It allowed me to reveal how the ongoing build-out of a national broadband Internet

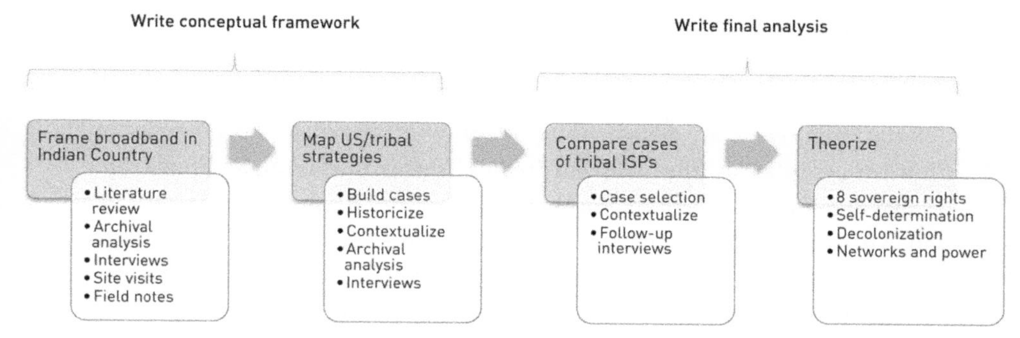

FIGURE 2.1. As a decolonizing methodology, reframing requires setting up conceptual space in which to investigate a challenge and provide a theoretical explanation that centers Indigenous experiences. Investigating tribal ICTs called for a recursive approach, beginning with (1) framing broadband in Indian Country by identifying relevant cases through interviews and site visits; (2) distinguishing tribal Internet infrastructural deployments through contextualized case studies; (3) conducting cross-case comparisons to identify high-level similarities and differences; and (4) theorizing the impacts of broadband Internet in Indian Country through Native American and Indigenous studies theories of sovereignty, self-determination, and decolonization, in addition to information scientific explanations of networks and social power.

infrastructure depends on the participation of sovereign Native nations. It also allowed me to understand broadband Internet infrastructures as a technology integral to the flourishing of Native peoples.

As a methodology, reframing clears the muddy waters. The temptation of many research reports has been to depict Native Americans as impoverished ethnic minorities within an expansive and wealthy United States. Supposedly, connectivity would be possible, under the right conditions: sustained local capital, sufficient demand, technical know-how, nearness to urban centers, and the adoption of the values of a technological society. References to the "smoldering conflict" of reservation border towns and the "cultural differences" that seemed to blockade technical advancement are common. So, too, are the many references to "bridging" the Digital Divide, as if widespread adoption of ICTs could somehow ameliorate the chasm of political and social marginalization of Native peoples effected through overlapping waves of preindustrial and industrial-era colonization of Native lands, waters, and bodies. Understanding the uses of ICTs in specific geopolitical contexts means understanding the character and limitations of the digital system as well as the will of the people in their historical moment.[15] In short, in 2010, what I could see was two black boxes: the black box of technology colliding with the black box of culture.

Attempts to describe Native uses of ICTs that were based on reductive scientific inquiry—surveys, polls, census data—could not clarify the historical and philosophical contexts shaping uses of ICTs in specific Native communities. These reductive study designs were not drawing out what Subcomandante Marcos described as "the hand that dreams," that is, the visions and goals that tribal and Native leaders had for their communities when deciding to utilize specific kinds of ICTs. Further, these studies were not nesting those visions and goals within a greater understanding of Native self-determination, a particular formulation of self-determination that harks back to distinctively Native philosophies toward harmonious and sustainable ways of life anchored in ancient tribal traditions. All these nuances were thrown into the black box called "culture," using urban, white, first-world techno-scientific settings as the reference point for an "Internet-ready" culture.

Additionally, there are general treatments of technology, media, or the Internet in Native and Indigenous studies that do not nuance the specificity of various technological forms and modalities. The Internet is a complex open system, and studying aspects of the Internet requires a statement about what specific facet of this complex open system is up for examination. In the past decade, researchers in Internet studies, computer science, information science, and media studies have published a number of books and articles about methodologies for investigating specific aspects of the Internet and digital systems. One need not be a software engineer or computer programmer to study social aspects of the Internet, but respecting the design process and understanding how systems and interfaces work go a long way toward unpacking the black box of technology. This also helps one resist the temptation to depict the Internet as a single hegemonic force threatening Native ways of life, a rhetorical argument that in a single turn obfuscates both the complexity of the Internet—or any digital technologies for that matter—and the complexity of Native ways of knowing and being. Likewise, there is another temptation to depict social media as a panacea for the marginalization of oppressed and voiceless peoples. This is an inversion of the first argument. Both arguments point to the values of a society that believes in progress through techno-scientific advance as well as the myth of the disappearing noble savage.

This is not to say that the Internet is purely utilitarian, a tool as simple and pure in its innovation as a hammer or a cup for holding water. Conceptually, it represents a richly featured terrain, with the topographies of technical networks shaped by a mixture of the personal agendas, political will, mundane habits, and desires of the individuals who create and use them. By the mid-

1990s, communications theorist Manual Castells predicted that ICTs would initiate dramatic societal shifts in a network society.[16] He wrote that place-based identity groups would strengthen their niche holds in their countries, utilizing web-based ICTs—e-mail, streaming video, chat—to communicate their plight and interact with like-minded groups in other parts of the world. He referenced the Ejército Zapatista de Liberación Nacional and its use of ICTs to broadcast news of its politically exigent status around the world.[17] Castells's acknowledgment of this choice is important for Indigenous scholars to note because it captures a factor that many of the technical government-funded surveys of US Native uses of ICTs were not, and that has to do with the political will of the people. In short, cultural differences are not an impediment to Indigenous peoples' decisions to implement ICTs. Instead, the political motivation for a people or a tribe to import or design and implement these tools varies based on local needs. In Castells's characterization, an organized Indigenous peoples movement was assuming command of a highly technical system from within a particularly sticky state of political exigency: ongoing colonization.

One can deepen this level of explanation. Sociotechnical theorists conceptualize ICTs as manifestations of human relationships, complex interfaces oriented toward satisfying the human will to connect and communicate both informally and across institutions. It is possible to consider institutions as de Landa describes them, crystallizations of human relationships. Systems of ICTs are thus concerted fabrications, many times bearing emergent properties that information scientists and scholars of science, technology, and society busily seek to discern. Culture is also about human relationships. Thus, thinkers positing cultural differences as a major reason for the lack of widespread uptake of ICTs in a particular community are really saying that they do not understand the nature of the relationships shaping the conditions for innovation in that community. In Indian Country, these conditions are many. There are layers of political, legal, corporate, tribal, family, and ancestral relationships, as well as relationships with mountains, bodies of water, deserts, plains, mesas, animal beings, and sky beings. There are relationships with bodies of thought, including Western, Indigenous, and local tribal philosophies, not to mention histories and practices. All these relationships are power-laden.

Indigenous scholars have begun mapping the power relationships between human and nonhuman actors—a poststructuralist approach—in Indigenous contexts.[18] There are different poststructural approaches. With regard to sociotechnical systems, Wiebe Bijker, Bruno Latour, John Law, and others utilize actor network theory to map the power relationships around systems by creating equation-like chains of human and nonhuman actors defusing conflicts toward achieving shared goals.[19] Actor network theory scholars often frame devices or components of a greater technological system as the nonhuman actors. Portraying humans and nonhuman technical devices as complementary actors in a greater system reveals moments when humans work in pleasant power-sharing accord with the devices in their lives and leverage devices to defuse a perceived threat to a work goal. Consider the irony of scholars ranting about the hegemony of the Internet as a force of colonial domination yet simultaneously depending on e-mail, Facebook, and smartphones to keep up with friends, relatives, and colleagues. The critique of particular dominating relationships around specific uses of systems built on top of the Internet is warranted, as is the awareness that in some parts of the world, activities on any given day are likewise supported or alleviated through specific uses of digital systems. In this web of relationships lie the factors shaping Native people's decisions about how to use ICTs and toward what end. At this point, the only way to understand this array of phenomena is to apply decolonizing methodologies to science, technology, and society studies approaches. We have to weave the impacts of colonial legacies into contemporary theoretical, practical, and policy-based characterizations of technological solutions in Indian Country. To avoid the very real impact of particular colonial mechanisms and Indigenous praxis in our built and lived environments is to obscure the potential for meaningful change in Native and Indigenous contexts.

To many in the wired and hyper-connected world, it is unimaginable and in many ways undesirable to live in a remote reservation with no landline phone service, slow Internet only in the public library fifty miles away, and an intermittent wireless signal. Throughout the day, an iPhone might be more useful as a weight to keep the napkins from flying off the kitchen table. Yet there is innovation and great creativity in these places where a tribal people can continue to stand strong as an independent-minded Indigenous people against modern state, federal, and epistemological encroachments and the urgent press to forget the land and the language and assimilate. Indigenous people dream with or without the pencil, the mobile phone, the social media account, or the design software.[20] But when we dream of the utility of these systems, for what purposes do we dream?

CHAPTER 3

The Overlap between Technology and Sovereignty

> At the very foundations of the world in which we live it is a unified world and cannot be reduced by techniques and rationality. Where traditional Indians and modern science are quite different is in what they do with their knowledge after they have obtained it. Traditional people preserve the whole vision, whereas scientists generally reduce the experience to its alleged constituent parts and inherent principles.
>
> —Vine Deloria, Jr., "Traditional Technology"

IN THE SUMMER OF 2011, I COMMENCED AN EXPLORATORY QUALITATIVE study into tribally centered information and communication technologies projects. I sought interviews with people working with or developing digital information systems designed to support the exercise or enforcement of tribal sovereign rights. My goal was to articulate instances in which ICTs and sovereignty interrelate within the boundaries demarcating Indian Country.

ICTs are digital devices that function as part of a larger system of people and devices to circulate information essential to the integrity of the hosting institution or organization. I conceptualized landscapes—and especially urban landscapes—as laden with invisible interconnected and at times disjointed systems of digital devices transmitting continuous streams of data and information from one server to another.

I applied a fairly open definition of data and information flows in the context of tribal sovereignty. Around nine months earlier, I had been working on articulating the significance of information for tribal governments with my colleagues in the University of Washington's Indigenous Information Research Group. From an operations standpoint, tribal governments are departmentalized into units, including health services, land management, education, member enrollment, law enforcement, and so on. Each unit has systems for sharing information with the others, with institutional partners, and with the federal agencies that support operations through grants and loans. For example, a tribal clinic may build information systems for reporting local statistics to Indian Health Services and the Centers for Disease Control and Prevention, as well as to the tribal council for the purposes of informed decision making. Our research group had been conceptualizing phenomena associated with the obstruction of information flows that are essential for the governance of a tribe, including when federal authorities or other partners misinterpret, misuse, or harness information for the purpose of exploiting tribal governments.

In one well-known example, the Havasupai Tribe partnered with researchers at Arizona State University in Phoenix to track the incidence of diabetes among the Havasupai people. Study participants donated blood samples, with the understanding that the researchers were looking for genetic markers for diabetes. But the researchers had a different agenda in mind and began testing the samples for incidence of mental illness and inbreeding. Operating within a frame of biological determinism, they asserted that the blood showed that the Havasupai people were not entirely Havasupai. Treating the blood as pure information—removed from its context, devoid of significance beyond that of the university lab—the researchers objectified the samples and invested them with values far removed from the desert canyon philosophy of the Havasupai people. This example helps in considering what Vine Deloria, Jr., wrote in 1999, that there is a difference between what traditional Indians and modern scientists do with knowledge after they obtain it, with modern scientists divorcing data and information from the unified world from whence it came, the originating episteme.[1] Worse, the ways of thinking that shaped the researchers' interpretation of the test results bore a colonial mind-set, with the Indians depicted as socially inferior and unwell while the purportedly technologically superior researchers gained credit for their advancement of genetic science.

As Indigenous information scientists, we recognized what had happened, how a people's blood had been reclassified as information and how that reclassification allowed the researchers to treat the Havasupai people with inhumanity. We also recognized that the cultural sovereignty of the Havasupai people—that is, the reality of their existence as a self-governing Native peoples free to live by ways of knowing developed over millennia within the ecologies of their homeland—would ultimately overpower whatever ill-educated results the researchers had prepared.[2] Indeed, people within the Havasupai community filed a lawsuit against the Arizona Board of Regents and the professor who manipulated the DNA samples.[3] Tribal people spoke to journalists about the mistreatment they had experienced. In Indian Country, the researchers were reproached for their breach of research ethics.[4]

Those of us in the Indigenous Information Research Group began considering how to convey to tribal leaders the importance of protecting tribal peoples' data and information as a matter of the integrity of tribal ways of knowing and modes of self-governance. Interpreting tribal sovereignty from a protectionist stance, we began considering how the political and legal sovereign rights of tribes, centered around cultural sovereignty, might be leveraged to protect against the misuse of tribal data and information.

At its most minimal, tribal sovereignty may be understood as the dynamic relationship between the will of a people to live by the ways of knowing they have cultivated over millennia within a homeland and the legal and political rights they have negotiated with the occupying federal government. Others have distinguished these as cultural sovereignty and legal-political sovereignty. At present, federally recognized tribes within the boundaries of the United States exercise the following eight rights as sovereign governments: the rights to self-govern, determine citizenship, and administer justice; the rights to regulate domestic relations, property inheritance, taxation, and the conduct of federal employees; and the right to sovereign immunity.[5]

For tribes, sovereignty refers to the integrity of a people, as well as the integrity of their government. It is important to distinguish between the two, because at present many Native and Indigenous peoples live under an imposed and therefore negotiated form of government, in which there is a clear memory of how Indigenous modes of self-governance differed from the colonial form of government. A free and autonomous Native people retain this memory by sharing information among themselves and with neighbors, which strengthens their knowledge of their homeland, shared history, Native language, ceremonial cycle, and lineage.[6] In this sense, the peoples' will to strengthen their inherent cultural sovereignty—through learning their tribal histories, languages, philosophies, spiritualities, and relationships with landscapes and sustaining healthy tribal families—can occur alongside whatever legal or political struggles the tribal government endures. The leaders of a sovereign tribal government also share information among themselves and with the leaders of neighboring governments to strengthen the tribal capacity for self-governing, determining citizenship, administering justice, and so forth.[7]

We knew that information sharing is integral to Native peoples and tribal government leaders, but we did not understand precisely how information and sovereignty interrelate. Specifically, I did not realize how completely tribal sovereignty shapes daily work in Indian Country and how ICTs play an integral role in circulating information critical to the daily exercise of sovereignty.

That summer, I drove from Tucson, to Phoenix, to San Diego and conducted phone and in-person interviews with nine individuals working on a range of projects, from tribal radio stations to oral history websites, law enforcement information-sharing centers, databases for tribal governance practices, tribal broadband policy making, and network certification programs. My goal was to sensitize myself to the dimensions of the interaction between exercises of sovereignty and uses of ICTs. I was trying to get a sense for what Vine Deloria, Jr., called the "whole vision," the underlying fabric and purpose motivating and giving meaning to Native peoples' uses of ICTs.

KPYT-LPFM: THE OPERATIONS BEHIND "THE VOICE OF THE PASCUA YAQUI TRIBE"

Hector Youtsey, the manager at the Pascua Yaqui tribal radio station KPYT-LPFM, which had just set up a streaming radio program, was the first to be interviewed. The station is housed in the old smoke shop, an adobe-style building beside the tribal casino about twelve miles south of the desert city of Tucson. The station placard bears the turquoise and red colors of the Pascua Yaqui flag, with the black-and-white outline of a radio tower pointing to the sky. We enjoyed a conversation about the beginnings of KPYT-LPFM in his office between the media and live recording studios. While we spoke, a deejay was helping the tribal higher-education director's son listen to his voice recorded live on the air for the first time. The station technician, a retired engineer, sat at a table in the bright sunlight, modifying an antenna for greater reception. Gesturing at a server rack, I asked Youtsey what it had taken to get the streaming radio program up and going.

Youtsey described his experience working for a commercial radio station in Tucson, and how, after a while, he became more interested in working for community radio station, where he could tailor the music and programming to community interests. He mentioned this to one of his friends, who was a councilman for the tribe. For a few years, the council members had been discussing how to get a tribal radio station going, especially to promote Yaqui-language programming and music and cover local news and events. Youtsey's friend asked him if he would be open to helping the tribe set up its station. This story about the start of the tribal radio station is an example of tribal leaders recognizing the need for community-level information that will strengthen the people's ways of knowing and community cohesiveness.

As it turned out, Youtsey was the right man for the job. His experience working with commercial radio regulations and community radio needs helped him take charge of balancing the Federal Communications Commission (FCC) operations standards and the requirements of the Pascua Yaqui tribal government. He set up the station by regularly updating the council members and also by developing relationships with the different tribal departments that were helping with the setup, from construction to provision of information technology (IT) services to the tribal library. He hired and trained tribal members to work as station employees and turned to his circle of radio colleagues and community radio advocates for advice and assistance on training and technical fixes.

Working in this way, he connected with Traci Morris and Loris Taylor of Native Public Media, an Arizona-based media advocacy nonprofit, and was able to advocate with the FCC for establishing a tribal priority in licensing radio spectrum that would fit the shape of reservation lands. In the past, tribes had difficulty acquiring licenses because the FCC was allocating licenses for cubes of airwaves over squares of land. When tribes would apply for access to airwaves above tribal lands, which are not in the shape of squares, they would find that competing radio stations already had licenses on or near tribal lands, effectively blocking tribes from using radio as a means of communicating local information to their communities.[8] In the end, the Pascua Yaqui Tribe ended up acquiring a low-power frequency modulation, or LPFM, license.

The official reservation lands for the Pascua Yaqui Tribe consist of 202 acres southwest of Tucson, but the more than eight thousand members of the Pascua Yaqui Tribe actually inhabit several barrios, camps, towns, and villages in and around Tucson and Phoenix and also live in family units throughout California, New Mexico, Texas, and other parts of the United States. As a people, Yaquis have resided for millennia throughout what is now northwestern Mexico and the southwestern United States. The original sacred homelands of the Yaqui people are located outside Guaymas, in the Mexican state of Sonora. In one of many violent confrontations with the Mexican state, in the late twentieth century, Mexican president Porfirio Diaz enacted a policy of capture and enslavement of Yaqui people defending their homelands or providing care to those Yaquis suspected of rebellion against Mexican federal or state authorities. Yaqui people were packed into trains and sent to work on hemp and sisal plantations in the Yucatán and Quintana Roo, far to the southeast of their homelands.[9] To this day, Yaqui families reside throughout both the United States and Mexico and share information about changes in their communities and neighboring cities and how federal, state, and tribal policies affect the health, spirituality, and well-being of the people as a whole.

While a low-power FM station serves the needs of people living on the reservation near south Tucson, the bandwidth is insufficient for meeting the needs of tribal people living throughout the United States and Mexico. The streaming radio station allows anyone living beyond the reach of KYPT-LPFM 100.3 to visit the tribal website and listen to language lessons, music, news, and other special programs. Youtsey worked with the tribal council and specialists in the tribal IT department to set up and test the streaming radio system. Shortly after setting it up, he began receiving e-mails and phone calls from listeners in unexpected places, thanking the station for the interesting programming and local music. Musicians submitted their CDs for radio play. Youtsey made sure that the deejays promoted community programs on air within half a day of receiving requests. The station technician began testing ways of bending the antennae so that the signal could be boosted through a technical modification in spite of the low-power designation. He organized a volunteer program that would teach youth about working in a radio station, creating programs, and recording and playing their own media on the air. It soon became clear that this theme of teaching and training tribal youth would pop up in every tribal ICT venture I observed.

Indeed, the individuals I interviewed during that summer would echo many of Youtsey's experiences utilizing ICTs to convey information for tribal community needs. Ideas for projects began with tribal leaders discussing the need for quality local information and then finding talented and experienced individuals among their networks of friends, family, and associates who could implement their ideas. These individuals would work as champions and managers of the project. In Youtsey's case, he champions the potential for community radio within the tribe, connecting local needs with the capacity of the technology. He advocates for tribal radio in local and national forums. He also manages the daily functioning of the radio station. This blend of activity—a form of ICT leadership—requires knowledge of the tribal community's history and geopolitical status, awareness of contemporary community needs and interests, an understanding of the policy and technical requirements needed to run the ICT project, entrepreneurial acumen, managerial skill, and a long-term vision for what the ICTs in question can do to improve community well-being. Over and over, I saw that strong relationships were key in acquiring capital to fund projects, developing technical training programs, acquiring hardware and software, hiring the right people for the right jobs, and advocating for needed policy changes with governmental agencies, such as the FCC.

In this case, I could see that the Pascua Yaqui Tribe's work toward establishing a radio station was also about increasing tribal community members' awareness of events, services, programs, and local news of interest to the tribe. It was about increasing the tribal government's ability to disseminate information to tribal members—information that could increase the tribe's capacity to self-govern—and it was also about creating a space where tribal members could advertise their own programs and events, such as back-to-school events and concerts, and share shout-outs to loved ones and, of course, the elders' words. Thinking back to expressions of cultural sovereignty, I could see how the radio station was allowing for the sharing of elder's knowledge through Yaqui-language programming and was cultivating young people's knowledge through workshops and educational and arts events. Thinking back to the mechanisms of colonization, I could see that the tribal radio station was serving an important decolonizing function by centering information flows within the tribal community, divesting non-tribal corporate ownership of radio spectrum over tribal lands, and retuning the use of the spectrum to promote tribal community events rather than mass media ClearChannel programming.

SMART WALLS AND TWO-WAY RADIOS: ICTS ACROSS THE TOHONO O'ODHAM NATION

After visiting with Youtsey, I spent time speaking with Police Chief Joseph Delgado at the Tohono O'odham Police Department. Like the Yaqui people, the Tohono O'odham people are both binational and transnational, having lived for millennia in desert and coastal homelands stretching from what is now northwestern Mexico through the southwestern United States. As a federally recognized US tribe, the Tohono O'odham Nation comprises more than 4,500 square miles of land located south of Tucson along the US-Mexico border. Indeed, the nation's southern boundary is also the US-Mexico border, a borderline negotiated through the 1854 Gadsden Purchase, when US ambassador James Gadsden sought completion of a southernmost US transcontinental railroad line, as well as reconciliation of outstanding property and citizenship claims made by American and Mexican settlers during the 1848 Treaty of Guadalupe Hidalgo. More than 150 years later, the US-Mexico border continues to be a contested space regarding landownership and access, as well as citizenship rights. The border is a testing ground for the national sovereign powers of the United States and Mexico, particularly with regard to economic expansion, nationalist identities, and the enactment of policies delimiting the lives and labor potential of all those who cross border checkpoints for work and family. Surveillance technologies—drones, identification with embedded microchips, infrared long-range cameras, elaborate checkpoints, interrogation techniques, and vehicle X-rays—pervade the region for hundreds of miles on both sides of the line. The challenges for Indigenous peoples residing in the US-Mexico border region are unique and complex. This is especially true for the Tohono O'odham people and their government, the Tohono O'odham Nation.

A few days earlier, I had traveled into the desert with Tohono O'odham human rights activist Mike Wilson, filling water tanks and leaving gallons of water for people without passports and green cards who cross illegally into the United States through the O'odham deserts rather than through border checkpoints. The Sonoran desert is harsh terrain, arid and rocky, reaching temperatures above one hundred degrees Fahrenheit during spring, summer, and fall and dropping to less than sixty degrees at night. Many people perish in these harsh conditions. A number of years ago, US Customs and Border Protection—formerly Immigration and Naturalization Services and now positioned under the Department of Homeland Security—designed a deterrence technique in which they positioned checkpoints at geographically temperate locations, thereby funneling people seeking to cross without papers through the harsher desert terrain. The goal of the program was to deter people from crossing.[10] Yet people still cross. Sadly, more people perish while trying to cross through the Tohono O'odham Nation than at all other points along the border.[11]

Wilson is critical of the Tohono O'odham Nation executive leadership for what he explains is their misreading and misuse of tribal sovereignty.[12] A US Marine veteran and a former pastor of a local Baptist church, he cautioned me about believing too much in the notion of tribal sovereignty. Born and raised on the US-Mexico border—internalizing it as a conflict zone for all who cross there—I empathized with his critique. Indeed, as a mode of governance, tribal sovereignty has its limitations.[13] While the sovereign rights of tribes continue to be the most powerful legal and political mechanism that US tribes have for negotiating with the federal government, and the top defense against state encroachment and private citizens wishing to profit off tribal lands and bodies, sovereign tribal governments can nevertheless also provide a haven for unscrupulous tribal politicians whose desire for power often overrides their compassion for humanity, the environment, and even their own relatives. Wilson is a deeply compassionate individual and identifies injustice where he sees it, which is, in itself, a Sisyphean labor given the often violent human and political dynamics of the border region. Sitting and working alongside Wilson reminded me quite a lot of visiting with my own relatives, cool-headed critical thinkers as familiar with the desert terrain as with the human dynamics that unfold in borderland emergency rooms and at the far edges of tribal ceremonial grounds, where those suffering psychosocial traumas unleash their troubles in ways that frighten all but the most grounded individuals.

I watched the changes in the beautiful desert landscape from the cab of Wilson's pickup as he drove us from one watering station to the next. From an information perspective, I sought evidence of telephone lines, radio towers, satellite dishes, wireless receivers, and the like. As we approached the border, we drove past a building that served as a base station for US Customs and Border Protection officers working on O'odham land. A large steel tower lay unused, in pieces, alongside the building. It was a smart wall tower, an expensive information system designed by Boeing about a decade earlier.[14] The goal of the smart wall had been to utilize 360-degree environmental sensors and wireless broadband technology to transmit data about movements in the landscape to roving unmanned aerial devices and back to border officers working at base stations and at strategic points in the field. Later, as we drove to another watering station, I noted old television sets, broken telephones, mattresses, and children's toys heaped beside a dumpster. I considered how tribal leaders must perceive the life cycle of devices—from design to deployment to recycling and elimination—within the taut geopolitical ecology of their homelands.

Questions of how the sovereign rights of tribes are tested at the boundaries of tribal lands were on my mind as I sat with Chief Joseph Delgado in his office across from the San Xavier Mission south of Tucson. Chief Delgado described how his officers undergo a critical decision-making process when they are alone out in the field and run across groups of individuals involved in illegal activity. There are parts of the Tohono O'odham Nation desert landscape where cell phones do not receive signals. Officers carry short-range radios as a communications backup. I asked about the systems they use to share information with authorities from other law enforcement agencies, such as the US Customs and Border Protection officers, the neighboring Pima County Sheriff's Department, and the Tucson Police Department. Chief Delgado described the fusion centers project sponsored by the US Department of Justice.

Fusion centers are strategically located organizations that take in and collocate information from state, municipal, tribal, federal, and other law enforcement agencies for the purposes of intelligence analysis. Chief Delgado referenced the infamous case *Oliphant v. Suquamish*, in which US Supreme Court Justice William Rehnquist decided that tribal courts could not try non-Indians residing on Indian reservations.[15] The number of non-Indian criminal suspects living on reservations is high. I considered what I know about the way crime and violence regenerate in the US-Mexico border zone and triangulated to consider the US-Mexico–Tohono O'odham Nation border zone.[16] Truly, Native Americans are border crossers. It is central to the Native experience, to exist as kin to an Indigenous people and yet to also exist as a marginalized subject of a dominant colonial government, a member of a tribe, and a voting citizen of a state and a federal government. From an information perspective, I thought about the asymmetries in information sharing that occur as tribes seek to make their information systems operable with neighboring municipal, county, state, and federal authorities, the trust that must be involved in making information-sharing decisions, and law enforcement consideration for public safety needs and the rights of tribal members and non-Indians living on reservation lands.[17]

The Tohono O'odham Nation hosts three casinos within the boundaries of its reservation. Chief Delgado described the work his team does there, watching for criminal activity associated with gaming operations and maintaining public order. With such a large and institutionally diverse landscape to monitor, Chief Delgado's officers work to uphold public safety in some places laden with robust ICT infrastructure and information flows, such as near the casinos and townships, and in other places thick with linguistic differences, no cellular and radio service, and regulations obstructing or curbing critical information sharing, such as at the borderlines and deep in the desert. Each year, during certain seasons, many people in the region, including O'odham people, Yaqui people, Mexican Americans, and others, undertake arduous pilgrimages from one mission to another, to family homes, and to other sites of prayer located alongside centuries-old routes from southern Arizona into the Mexican state of Sonora. These pilgrimages are an important aspect of Tohono O'odham spiritual practice and history. Chief Delgado described a communications technique that the public safety officers employ to alert people on pilgrimage about points of safe passage, sudden thunderstorms, and fire warnings. Listening to Chief Delgado, it became clear that dispatch centers, fax machines, cellular phones, shortwave radios, Facebook pages, and tribal radio stations playing through the speakers of four-wheel-drive trucks ranging through the desert are all part of a flexible system of devices for sharing information critical to maintaining public safety in the remote parts of Indian Country.

It is clearly a challenging task, designing information systems that are based on the same regulated technical standards but strictly curb flows of information according to the rights of individuals, the needs of institutions and community groups, and national and tribal governmental policies. While technically sophisticated, and certainly costly, the smart wall lay in pieces across the desert, revealing both the limitations of industrial materials in harsh conditions and the limitations of complex systems imported into regions where the jurisdictional tensions of multiple sovereign authorities preclude easy intra-institutional collaboration. Maintaining public safety for the Tohono O'odham people—challenging in any scenario—requires that the Tohono O'odham Police Department innovate as best it can, given the social, technical, and political constraints of its jurisdiction. In some situations, four-wheel-drive trucks, two-way radios, and cool heads can accomplish more than a smart wall. Speaking with Chief Delgado showed me that a keen understanding of the tribal landscape and communities therein provides the foundation for designing a sociotechnical assemblage that serves a tribe, in particular a sovereign tribe standing strong in the midst of geopolitically contested terrain.

In a month's time, I spoke with more individuals about their projects: Richard Alum Davis of KUYI Hopi Radio; Joan Timeche of the Native Nations Institute; Sandy Littletree of the Knowledge River Tribal Librarians Oral History Project; and Traci Morris of Native Public Media. With each person I interviewed, I learned more about how uses of ICTs relate to exercises of tribal sovereignty.

KUYI HOPI RADIO: PROGRAMMING TO MATCH THE RHYTHM OF HOMELAND

Richard Alum Davis, the station manager at KUYI Hopi Radio, described how the Hopi Tribe set up its own community radio station. The Hopi people are a Pueblo people who have resided for millennia in the canyon and desert mesas in what is now the Four Corners area, where New Mexico, Arizona, Colorado, and Utah meet. The Hopi Tribe reservation land is presently located in northeastern Arizona—surrounded by the Navajo Nation reservation land—and the tribal government serves Hopi people and Tewa people residing within its borders. Bearing a complex philosophy and spiritual practice, Hopi people have for generations exercised a communal mode of self-governance deeply rooted in the seasonal rhythms of their homeland. As members of a federally recognized tribe, they have also developed a government that interfaces with the Bureau of Indian Affairs and other federal agencies.

When Davis and I first arranged for a phone conversation, I laughed because we both had to plan to park ourselves in unusual locations where we could receive cellular phone signals. He was heading to a parking lot near a gas station in Hopi where he could receive a signal, and I was sitting in the back of my father's truck in Mesilla, New Mexico, facing northwest. I'd been to Hopi before to visit friends and was soothed by the blue sky filled with traveling rain clouds, the subtle shapes of the windswept desert floor, and the striking rock mesas. I had seen a hawk dive full speed down the side of a mesa, hunting from cool clear sky to heated rock wall.

It is difficult to express appreciation for the ecology of a landscape to others. It takes a great deal of deep listening and working within a tribal community to begin to experience the seasonal rhythms in one's bones and to understand the reason for adapting to those rhythms. Davis described how deejays at KUYI play certain kinds of music at certain times of the year, attuned to the meanings of the seasons. They avoid edgy or aggressive music during the gentle winter months. At other times, deejays select music from other Native peoples, encouraging the local community to open their ears to new sounds from peoples who likewise understand what it is to live in right relation with a landscape. Language-learning opportunities are included as much as possible. KUYI personnel seriously discussed the pros and cons of airing tribal council debates during election season. On the one hand, the radio could provide critical election information to community members, especially homebound elders, who could not attend the debates in person. On the other hand, that kind of self-governance information is a private matter for tribes. Messages heard over the radio—separated from body language, context, and visual cues—could be misinterpreted or misunderstood. The radio station did not want to be perceived as "airing the dirty laundry" of internal council matters.

Davis's explanations reminded me of Hector Youtsey's decisions to train his deejays not to play certain kinds of Yaqui music at certain times during the yearlong ceremonial cycle. Adapting ICTs to the ecology and internal rhythms of tribal homelands requires respect for language, ways of knowing, tribal privacy and security, and modes of self-governance. The Indigenous Information Research Group had been considering this dimension as one of those that most distinguish Native uses of ICTs: in many Native communities, certain kinds of content—especially content that is sacred in nature and that threatens the security of private tribal self-governance operations—may not be recorded and broadcast across any form of media. In many Native communities, cellular phones and recording devices of any kind, including sketchpads, cameras, and audio and video recorders, are prohibited on ceremonial grounds, especially during moments of prayer.[18] Our group's discussions of this issue contributed to my colleague Miranda Belarde-Lewis's investigation of YouTube as a space for sharing videos of sacred and social Native dances.[19]

I began to think about how the notion of access must differ for Native peoples, who must contend not only with the poor-quality content that exists about Native peoples but also with the policies and geography of their reservation, as well as those of the surrounding tribal, municipal, county, state, and federal governments. The FCC decision to adjust spectrum licensing to fit the shape of reservations, and not just the shape of a block of cubic miles, goes a long way toward giving a tribe access to the AM/FM radio spectrum coursing through their homelands. It is up to the project personnel to decide how to make appropriate use of that spectrum within the geopolitical constraints of the reservation. Decisions about how to use systems to disseminate information within tribal governments and communities point to the tribes' rights to a mode of self-government in which the people within the tribal community have to discuss for a long time the ethical, pragmatic, spiritual, social, and legal considerations around the sharing of knowledge and information by technical means. The pervasiveness of colonial mechanisms for turning information and knowledge about Native peoples against them is so common and expected that questions of access, security, disclosure, safety, privacy, and rights to privileged knowledge are central for tribal communities, particularly as related to the tribal regulation of domestic affairs and administration of justice and the peoples' work sustaining ceremonial cycles, healthy families, and ancient tribal philosophical and spiritual practices.

THE NATIVE NATIONS INSTITUTE AND KNOWLEDGE RIVER: ICTS FOR SHARING KNOWLEDGE

Acquiring the devices and setting up a system for sharing quality information is only one step in the process of implementing ICTs for a tribal community. The need for quality information within a Native or tribal community first drives the decision to utilize ICTs. This became clear to me as I spoke with Joan Timeche, director of the Native Nations Institute located at the University of Arizona in Tucson. Since 2001, the Native Nations Institute has served as a research and policy institute focusing on issues of self-determination, self-governance, and economic development for tribes. One of its main goals is to disseminate research results, policy implications, and lessons from leaders in Indian Country back to tribal leaders for purposes of informed decision making. The Native Nations Institute leaders participate each year in the Honoring Nations award program through the Harvard Institute on American Indian Economic Development. When we met, Joan handed me a copy of the past year's Honoring Nations program. I scanned the booklet and quickly noted that a majority of the award-winning projects were focused on building information systems to circulate quality information specifically for the purposes of upholding the operations of sovereign tribal governments.

The Mille Lacs Band of Ojibwe of the Great Lakes region utilized data about the local ecology in writing the Minnesota 1837 Ceded Territory Conservation Code regulating subsistence hunting and fishing. The code has been incorporated into regional district and appeals courts, resulting in increased understanding between tribal members and non-tribal neighbors who hunt and fish in the same terrain. Similarly, the Coquille Indian Tribe worked with the Smithsonian Institution and the University of Oregon to design the Southwest Oregon Research Project, an archive of cultural, historical, and linguistic documents pertaining to the tribal peoples of the area. Copies of documents were given to regional tribes during potlatches, contributing to a regional restoration of knowledge of Native peoples. Leaders in the Gila River Tribe needed a way of providing affordable and reliable phone service for their people residing on the reservation in southern Arizona. They started Gila River Telecommunications, Inc., a regional phone and Internet service provider for tribal residents and neighbors. Different tribes use geographic information systems to keep track of wildlife, water quality, and land uses for tribal land management. Tribes create systems for protecting pottery, weavings, petroglyphs, sacred dances, and artworks and aligning tribal, state, and federal policies in this regard. Tribes utilize ICTs toward language revitalization, including the above-mentioned radio stations, online learning modules, and digital storytelling tools. Almost all the honorees listed in the Honoring Nations program were also concerned with preserving lands for youth and educating future generations.

Sandy Littletree, then with the University of Arizona Knowledge River Program, worked with Latino and Native American students seeking a degree in librarianship. Faced with a scarcity of literature and in need of a way to teach students about their chosen profession, Littletree partnered with friends and colleagues in the American Indian Library Association and the

New Mexico Tribal Libraries Foundation. They filmed longtime tribal librarians speaking about their experiences and posted the videos in an online oral history archive. At the time of my visit, Native Nations Institute personnel were also preparing to launch a subscription database of video lectures by leaders in Indian Country speaking on a range of matters pertinent to dimensions of tribal self-governance. Intended to do more than collect data, these and the aforementioned information systems were designed to pass on Native leaders' ways of knowing.

Speaking with Joan helped me understand how tribes develop information systems for collecting local data that can be used for local decision making and building intergenerational knowledge. Inevitably, the decisions that tribal leaders make interface with the decisions and practices of neighboring governments. Of particular interest were those information systems designed specifically for intertribal and intergovernmental information sharing. But of greater interest were systems that focused on providing Internet access, as every individual I spoke to not only referenced the lack of quality information for tribal communities but also mentioned in passing the lack of basic phone, cellular, and Internet service in many tribal homes. Indeed, meeting with Traci Morris of Native Public Media and, later, with Matt Rantanen of the Southern California Tribal Digital Village Network highlighted the urgent need for reliable and affordable Internet service within reservation communities.

I had entered the field that summer understanding that information was important for the decision-making process of tribal leaders. I came to understand that the cultural sovereignty of a people relates to the ability of elders and experienced members to share ways of knowing with younger members. I saw how tribal geopolitics—political boundaries, physical geography, seasonal cycles, self-governance procedures—shapes uses of ICTs. Relationship building and partnerships are critical, as all the projects I learned about began with a few leaders sharing ideas and then tapping into their network of friends and colleagues, looking for individuals who could implement ICT projects. Project leaders possessed a unique skill set, capable of managing daily operations, advocating in local, state, and national arenas, and listening to and working with tribal leaders to articulate the long-term vision for the ICT project within the community. I could see how the content streaming across ICTs contributes to the local mode of self-governance, as political issues are debated across these channels. Project leaders continuously assess community needs and think about ways of applying technical know-how in order to meet them.

NATIVE PUBLIC MEDIA: BROADBAND INTERNET SHAPING CREATIVITY IN INDIAN COUNTRY

Traci Morris, then director of operations at Native Public Media, met me for an interview in a busy coffee shop near downtown Phoenix. As far as I could tell, each visitor to the coffee shop had a smartphone. This was a far cry from sitting in the back of my father's truck trying to receive a cellular signal near the Rio Grande, and far removed from the US-Mexico–Tohono O'odham Nation borderline, where the smart wall sits in pieces and an industrial lamp powered by a braying generator lights all who cross through the border fence at night. A longtime advocate for Native community radio, Traci was adamant about the impact that radio can have on Native communities. But she was more adamant about the impacts that broadband Internet can bring to Native communities. A regular media advocate in Washington, D.C., Traci assured me that people in Congress do not understand what it is like to be in a place with no cellular or landline phone service, such as in Indian Country. She described inviting a senator to visit a reservation and watching his body language as he realized he was getting no reception on his cellular phone and that if he was not getting reception, no one else was either. She also said many people do not quite understand the implications of broadband Internet for reshaping work and creativity in Indian Country. She described the digital dome at the Institute for American Indian Arts, a 360-degree digital recording space where students record Native dances and make films. What were the implications of this kind of technology with regard to the Native art of storytelling and other creative expressions? What kind of knowledge could be archived for future generations?

The smart wall is a broadband technology. The Tribal Librarian's Digital Oral History website runs at broadband speeds. The video lectures housed in the Native Nations Institute leadership database soak up a great deal of bandwidth. I wondered how many people in Hopi or in my own tribe have Internet speeds in their homes or workplaces that are fast enough for them to access this kind of content. During my fieldwork, my ability to convene with the Indigenous Information Research Group depended on driving to a café or a hotel with a connection fast enough to support videoconferencing. I wondered what it would take to give every tribal leader in Indian Country an affordable smartphone, tablet, and unlimited data plan.

Already attuned to the presence of digital devices, I began to conceptualize Indian Country as a vast expanse of geopolitically interrelated landscapes peopled by leaders sharing information about their tribes across a range of digital devices: smartphones, laptops, workstations connected to server rooms connected to broadband towers connected by fiber-optic cables to nodes buried alongside nearby interstate highways. There are dark spots in Indian Country, where no one receives any service due to the technical limitations of the devices. There are gray spots in Indian Country, where the elders have determined that no recording devices of any kind may be used out of respect for ceremonial rhythms and the sacred landscape. There are places in Indian Country that are extremely wired, where youngsters connect with one another on Xbox Live, grandmothers play the slots at the casinos, and young activists update anticolonial memes on their Facebook timelines. Previous studies positioned Native Americans as digital have-nots.[20] Through listening to the experiences of those working with ICTs in Indian Country, I saw that this was not the case; rather, like everything else that occurs within the boundaries of reservations, decisions about ICT infrastructure and uses must be negotiated across the local geopolitical and epistemic terrain.

Within my own ways of knowing, cultivated from growing up going to ceremony with my family in Old Pascua and running around the deserts surrounding Tucson and the river valley of Mesilla, New Mexico, I had come to see each moment as a blossoming, an unfolding within a greater dynamic of endless creation. I had read the writings of Vine Deloria, Jr., and Martin Heidegger alike on technology as a point of becoming, when human hands bring into being a system designed for the purposes of human expression.[21] But while Heidegger wrote about the technological domination of the natural landscape by a superior race of men, Deloria wrote about all human creativity as acts within this endless cosmic creation, an insight into which Native peoples bear a particular understanding by virtue of their spiritual relationships with the landscape and relation to all the beings therein.[22] ICTs in Indian Country serve purposes focused on Native peoples' expressions of their cultural sovereignty. Likewise, there are many examples of information systems in Indian Country designed for the purposes of supporting the operations of tribes. However, none of these can function without the availability of affordable and robust broadband Internet.

CHAPTER 4

Sociotechnical Landscapes

Tribal Broadband Deployment

The alternative point of view—the social shaping of technology—recognizes that technologies in general (and information technologies in particular, in our case) are the outcomes of social action. They are generated by people operating in social contexts (of all different sorts) and at particular historical moments, all of which shape the imagination of what needs technology might meet and in what settings it might be employed.

—Paul Dourish and Genevieve Bell, *Divining a Digital Future*

THOUGH THE EXPERIENCE OF USING WIRELESS DEVICES MAKES IT seem that the Internet is invisible, intangible, and placeless, the undergirding of the Internet is in fact quite visible, tangible, and place-based. In the winter of 2011, I returned to Seattle holding in my heart and mind, and in the assiduous notes of my research log, the narrative proof of what different Native and tribal leaders, educators, and activists were creating not only through the availability of affordable Internet access and networked technologies but also through networked thinking. Understanding the contemporary political will of various diverse Native peoples means conceptualizing the legacy of Indigenous histories in places, including the drive of many Native peoples as individuals to work together for social and political goals with or without federal or state government support. Likewise, understanding technology means conceptualizing digital objects not as just devices but rather as interfaces within a greater web of interconnected individuals, devices, and systems, in many ways upheld by the overarching work goals of preexisting institutions.

Leaders in Native and tribal communities harness information and communication technologies to accomplish information-sharing goals. The ways leaders collaborate depend on the individuals they know, and how those individuals can work together to accomplish mutually acceptable goals across institutions as diverse as distinct tribal governments, intertribal organizations, universities, nonprofit organizations, and federal funders. While there is great imagination about how ICTs "flatten" the divergences of time and space, thereby increasing the efficiency of information flows and work flows, for Native peoples, the realities of the jurisdictions and borders shaped by legal, political, and cultural sovereignty actually distinguish the spatial topography of ICT networks. These networks shape how Native peoples access and utilize various ICTs. Information is not "free" in Indian Country, and it certainly is not free-flowing; rather, it is shaped by the geopolitical and geophysical terrain, histories of colonization, linguistic choices, ceremonial cycles, protocols of respect, and values around sovereignty, revitalization, and tribal governance needs. All these boundaries and barriers shape Native peoples' choices and decisions around the design and integration of ICTs. Thus it is important to approach Indigenous digital endeavors as creative efforts to apply tools and techniques to addressing local needs and establishing a direction into a locally imagined future. Embracing this conceptual shift can help further unpack the black boxes of culture and technology, as well as center the lived experiences and political realities of Native peoples and the working conditions of tribes.

PLACE, INDIGENEITY, AND THE MATERIALITY OF THE INTERNET

When tribal leaders consider the build-out of any major infrastructure on tribal lands, they must take into consideration a number of factors relating to tribal codes, policies, community needs, financial capacity, and terrain. Land is precious in Indian Country, not merely as an asset or a development resource, but also as an epistemic topography of tribal ways of care and knowing. Native peoples' homelands have historical, spiritual, ecological, and political significance. It is perhaps easy for city-living folks to imagine the Internet as something "out there," as invisible and ephemeral as droplets of water in the air we breathe, and with data centers and network operations as nondescript as the next strip mall. It is more realistic for tribal residents to conceptualize the Internet as something "right here," with decisions about where to build towers shaped by seasonal rhythms of hunting, wildfires, and prayer, not to mention the matter of land and edifice allocation. Tribal communities take time to discuss appropriate uses of WiFi (wireless fidelity) and mobile phone use, in particular as these relate to the transmission of tribal community news and traditional knowledge. The Internet is made of quite tangible expensive material. The kinds of decisions that tribal ICT champions make to support the build-out of the Internet across tribal lands strike at the core of how tribes view the flow of information and knowledge within and around homelands, as well as how tribal leaders envision technology shaping modern tribal approaches to the exercise of sovereignty and self-determination.

At this point, it is helpful to explain one way of thinking about the components of a broadband Internet infrastructure. Generally, these consist of (1) the network system across which digital content is streamed; (2) the digital content itself; (3) the tangible devices that compose the network hardware, such as computers, cables, towers, and servers; and (4) the policies that regulate the build-out of the network, including uses and content flows. The network system, content, devices, and policies are created and managed by technicians working out of operational centers, such as university labs or network administrative offices. Location means a lot. Contractors must build roads to the towers and lay the foundations. The towers must be positioned over wide swaths of terrain—usually on hills, mountainsides, or other stable, flood-free high points—so that they can transmit signals to each other and down to buildings, be fire-proof and constantly cool, and be where network administrators can oversee massive flows of data. The whole architecture demands an uninterrupted power supply. At this point in the history of digital innovation, the broadband grid depends on the availability of a reliable electric grid.

Generally speaking, the term *broadband* refers to a digital communication channel of at least 256 kilobytes per second and operating in distinct contrast to earlier modes of single-channel dial-up. Technically speaking, *broadband* actually refers to the ability of a device to transmit multiple signals across multiple channels: fiber-optic cable, coaxial cable, and wireless, for example. At present, broadband Internet is usually transmitted by one of three technical setups: fiber-optic cable, wireless, or satellite networks. Robust networks include multiple delivery modes. Installing cable means burying miles of terrestrial cable or stringing aerial fiber-optic cable on poles. Setting up a wide area network for regional wireless delivery requires setting up towers, transmitters, and receivers. Satellite services bounce off orbiting satellite transmitters and introduce latency: the amount of time it takes for a packet to reach the satellite and travel back to the terrestrial receiver. When thinking about setting up Internet networks, systems designers conceptualize in four dimensions: what is underground, the layout of the visible terrain, how packets might travel through airwaves and, in some cases, across airspace, and all of this over measures of time.

Add to these (1) the encompassing market forces that determine Internet service supply and demand; (2) the physical geography shaping where and how towers, fiber-optic cables, satellites, and receivers are positioned; and (3) the political and institutional jurisdictions that shape the nature of policies and construction. There are also the people who support the build-out of broadband infrastructure networks. These include network administrators, content designers, policy experts, entrepreneurs, vendors and distributors of hardware and software, construction workers and contractors, industry and university researchers, lobbyists, system and interface designers, and Internet consumers of all kinds. Designing and building out a broadband Internet infrastructure requires a massive orchestration of individuals working through institutions. These individuals develop ways of building out the network across the policy and workplace constraints of their institutions, the physical terrain through which the infrastructure will be built, and the technical specifications of software and devices. The deployment of a broadband infrastructure is very much a place-based and institutionally supported enterprise.

Ecologically speaking, a broadband Internet network is a system that supports creative possibility for people who use the resulting services. The infrastructure of that network becomes the backbone for online forms of creative expression. While the people who advance the infrastructure become an integral part of the ecology of the entire system, the network itself becomes an integral part of the capacity of those who rely on Internet access to work and play in an online environment. Broadband network designers and advocates create the means for others to engage creatively in an online environment. This is why ICTs, and especially broadband Internet infrastructures, can be thought of as concerted fabrications. This is also why human-computer interaction researchers—specialists in designing interfaces that translate digital noise into useful human experience—now regard the design of systems as socially shaped, and neither purely technically determined nor entirely socially determined. Indeed, in their 2011 assessment of the future impacts of ubiquitous computing, informatics professor Paul Dourish and Intel researcher Genevieve Bell acknowledged that the diversity of people "operating in social contexts (of all different sorts) and at particular historical moments" promulgates and pursues various technological imaginaries.[1] Thus we can deduce that in different places within Indian Country, large-scale Internet infrastructure provides the backbone for the various kinds of online creative expressions that users develop and for the network administration work necessary to maintain the base layer of operations.

To the layperson, getting Internet access might be, conceptually, a matter of purchasing a laptop at the mall and calling AT&T or Comcast for a subscription. But the laptop is really only one device for plugging into an existing technical infrastructure, and the Comcast salesperson is only one individual working to support the whole enterprise. To understand the social and political impacts of ICTs means understanding the politics of the grid; focusing only on the device or interface—the mobile phone or website—replicates the fascination with the pyrotechnics. These "may hold our fascinated gaze, but they cannot provide any path to answering our moral questions."[2] What I have observed tribal ICT personnel do in their work is shift the conversation in policy-making arenas from pyrotechnics—talk of tribal websites and streaming radio programs, Facebook pages and digital libraries—to financial investment and political support of the undergirding broadband networks. In many ways, especially for tribes that are underserved or unserved by large telecommunications and Internet service providers, the tribal website and digital community archive are actually the realization of the often invisible efforts of tribal ICT personnel, many of whom worked for years advocating and negotiating for affordable phone and Internet service for their communities.

Unfortunately, at this point in history, it is difficult to depict the broadband landscape in Indian Country from a reductive viewpoint—figures and maps. At present, there are at least twenty tribally owned or Native-owned Internet service providers or in Indian Country, with more in the early establishment and development stages. There are no publicly available reliable data sets assessing Internet coverage in Indian Country, either in terms of technical reach of existing infrastructure or in actual numbers and locations of users, although some tribal service providers have these numbers as part of their private business strategy. As of this writing, affordable, reliable, and robust broadband Internet services continue to be scarce in remote reservation communities. Recent studies reveal that Native families in urban settings who have higher incomes and educational achievement are utilizing the Internet through mobile phones at a greater rate than previously expected. Basic Internet access through mobile phones is still quite different, however, from robust Internet access, the kind in which members of a household enjoy multiple strong Internet connections across multiple productivity devices.

While there are no exhaustive data sets at this point from which we can assess digital access, use, and connectivity across the diverse demographics of Indian Country, we can, as of this writing, still safely presume that robust Internet access and productive use of the Internet (as opposed to basic consumer uses) continue to be limited. We can base this presumption on the following: (1) ICT devices such as smartphones, laptops, tablets, and gaming consoles continue to be expensive for the average Native American household; (2) subscription rates for broadband cable, wireless, and satellite access continue to be more expensive for people residing in rural and remote locations, with many reservation communities located in such regions of the United States; (3) high unemployment rates in Indian Country mean there is less opportunity for individuals to gain Internet access through workplace computers; and (4) there are few regularly published data sets on numbers of users accessing Internet services through reservation schools, libraries, elders centers or computing centers, or nearby public schools and libraries; and (5) there are few studies that measure digital literacy skills and Internet uses, both of which represent different measures from those associated with basic technical connectivity.

Yet ICT devices are available and are being used in Native communities. In 2010, the White House released the National Broadband Plan, which included estimated figures on existing levels of coverage across the fifty states, and a general strategy for supporting infrastructural build-out such that each US home should have Internet coverage of speeds up to 100 megabits per second by the year 2020. However, Shana Barehand Green, former Federal Communications Commission employee and tribal telecommunications taxation expert, argues that while it looks like many parts of Indian Country are represented on the maps, what the maps really show are the specifications for the hardware, as reported by the telecommunications companies, were the hardware to function under optimal operating conditions. This means that the maps do not account for the limitations imposed by physical geography (wireless and satellite services are based for the most part on line-of-sight technologies), inclement weather, or regional monopolies fixing rates and blocking competition.

I had to smile. My family residing in southern New Mexico, for example, has never been able to maintain landline phone service because seasonal rainfall each winter and summer washes out the phone lines. When we first installed a wireless modem, we discovered that the signal would not penetrate the thick adobe walls of the family home, limiting the places where we could set up a desktop computer and laptops. During holiday visits home, when I need to take phone calls, I carry my smartphone outside and face northwest. Unfortunately, the touchscreen interface shuts down in summer temperatures over ninety degrees. My family does not live in a remote part of Indian Country; they are a fifteen-minute drive from an urban center. While the National Broadband Plan is intended to serve as a big-picture guide, in the end, the complexities of broadband Internet coverage, infrastructure, affordability, access, and use are not well represented in that single document or the accompanying maps and are not adequately represented at all within the diverse landscapes that make up Indian Country.

Meanwhile, to the north in Ontario, ICT champions in the Nishnawbe Aski Nation have been working for years to support broadband Internet access for First Nations peoples throughout Canada. The Kuhkenah network, or K-Net, began as a demonstration project formed out of a partnership between the six tribes of the Northern Chiefs Council and IndustryCanada. Since its inception, community journalists and researchers from McGill University, Montreal, and the University of Guelph have been documenting the build-out of this community-based multipoint network across the remote, densely forested lands of six tribes and the affiliated mountain and lake communities. The effectiveness of their partnerships has allowed K-Net leaders to advocate for the ICT and broadband needs of Native and Indigenous peoples through the Assembly of First Nations, the First Nations Technology Council, and the Indigenous Commission for Communication Technologies in the Americas. All K-Net documentation, including photos, videos, and plans, are posted online so that others may learn from K-Net's experiences as its leaders develop their own community broadband networks. First Nations in western Canada have access to a tribally owned satellite network, which is supplemented by a growing fiber-optic infrastructure. At present, the First Nations Technology Council is working on developing an integrated information management plan, as well as models for broadband networks and associated technology applications for all First Nations communities. Those working in tribal telecommunications policy must keep asking, What conditions shape this divergence between a cohesive First Nations broadband Internet strategy and the lack of a cohesive strategy for US tribes? How is federal funding for state and provincial broadband planning efforts shaping tribal planning efforts? How are private industry and small business shaping Internet accessibility in Indian Country? What are the many strategies tribes employ to acquire affordable Internet services for their communities?

These questions are important because they relate to fundamental questions about tribal sovereignty and self-determination around the build-out of a major communications infrastructure. Before we can answer these questions, though, it is essential understand how ICT champions build their own broadband Internet infrastructures for tribal communities across tribal lands.

PART I. TDVNET: CONNECTING THE NATIVE PEOPLES OF SOUTHERN CALIFORNIA

In August 2011, I drove to the Pala reservation outside San Diego, where I met with Matt Rantanen (Cree), TDVnet administrator and director of operations for Southern California Tribal Technologies (SCTT). Pala is northeast of downtown San Diego, higher up from the coastline, amid a range of boulder-strewn hills and valleys.

I drove south from Los Angeles, passing Marine Corps Base Camp Pendleton and two US-Mexico border checkpoints on the way. Sensitized to the geopolitical terrain of Indian Country, I noted this border enforcement as a continuation of three interrelated historical colonial legacies: the first being the suppression and forced reorganization of the coastal Native peoples—Kumeyaay, Tongva, Chumash, Cupeño, and others—into the mission system by Spanish, Mexican, and American settlers; the second being the US acquisition and subsequent occupation of northern Mexico under the Treaty of Guadalupe Hidalgo; and the third being the positioning of coastal California as a base of military operations to support twentieth-century US imperial expansion into Asia and Oceania. To my eyes, the development of information systems in Indian Country necessarily coincides with the development of US border enforcement systems.

Having read up on the history of southern California tribes, I learned that one of the first techniques US and Mexican colonial authorities utilized to prevent Native peoples from communicating regionally was damming waterways and preventing mission Indians from traveling via coastal and river routes. It is a history somewhat hidden in the published historical record, though apparent in the peoples' original names. Many of the peoples of what are now the southwestern United States and northwestern Mexico are named for flows of water. Knowledge was shared among diverse peoples along water routes: news, tools, stories, goods, medicines, maps, languages, and histories spread all the way from the southern continent to the far northern reaches of the Americas. The late nineteenth- and early twentieth-century remapping of the American western territories into states and cities coincided with the buildout of the transcontinental railroad, the diversion of major waterways, and the forced containment of Indians in the reservation system and, in California, the mission system. Many California Natives died during this turbulent time. American settlers kidnapped Native children and forced them into boarding schools. Many of the tribes of southern California now bear the names of missions, yet the memory of the peoples reaches back to an era that precedes the entry of early US and Mexican nationalist settlers.

Matt Rantanen, a Pala tribal administrator, shared with me the story of how the leaders of the Southern California Tribal Chairmen's Association (SCTCA) leveraged key partnerships with Hewlett-Packard engineers and researchers from the University of California, San Diego (UC San Diego), to acquire Internet access for the nineteen tribes bordering San Diego County. The idea of acquiring Internet for the tribes began in the mid-1990s, when tribal leaders realized that even though the reservations were located fairly close to an urban center, the peoples living in canyon, valley, mountain, and rural communities could not afford the expensive satellite access plans offered at the time. Many tribal residents also lacked phone lines. Regional Internet service providers explained that it was too expensive to build the infrastructure to reach these communities and that the demand would not produce sufficient profit. Later, I would discover that tribal broadband project leaders commonly offered this explanation for their inability to find a service provider for their remote, rural reservation communities.

Rantanen recalled that the peoples of the nineteen bands and tribes encompassed by the SCTCA were one people—political neighbors and blood relatives—before the imposition of the mission system and that TDVnet facilitates the communication that brings the people together again. He showed me a map and explained the reach in miles of the TDVnet backbone, with its primary towers located at high points overseeing the valley communities, and how the spectrum is allocated across each operational node. Spreading his fingers above each node, he asked me to imagine the array of pathways through which WiFi Internet service is provided to tribal administration buildings, schools, and libraries.

He said that in the late 1990s, at around the same time that SCTCA leaders were brainstorming how to acquire Internet access for the tribes, UC San Diego physicist Hans Werner-Braun was figuring out how to transmit astronomical data from satellites to the San Diego Supercomputer Center. Werner-Braun already had a group of engineers working on the project through the High Performance Wireless Research and Education Network (HPWREN), a broadband Internet network designated for scientific use. The engineers identified an optimal location within reservation lands for a tower that would receive and transmit satellite data. Werner-Braun approached the SCTCA, describing his plan to build-out the HPWREN backbone. He explained how the backbone could stream wireless signals to receivers in tribal homes and buildings.

The SCTCA identified a tribal member, IT specialist Michael Peralta, to meet with Werner-Braun and learn about his plan. It was a kitchen table meeting. Werner-Braun drew up a model for Peralta, showing him how to bounce signals from one room of the house to another using transmitters and receivers. Of course, the overarching concept was to provide high-speed wireless Internet service across towers set up on mountaintops and peaks, channeling spectrum down to valley administration buildings and residences.

SCTCA leaders found the concept worth an investment. They partnered with UC San Diego ethnic studies professor Ross Frank and drafted a proposal for a tribal broadband network through the Hewlett-Packard Digital Village program, including HPWREN and Hewlett-Packard engineers as key partners and technical consultants. The goal was to build a network that could sustain broadband operations across four domains—education, culture, economic development, and infrastructure—and to have tribal members administer the build-out, from design to implementation. In 2001, after receiving a three-year, $5 million grant from the Hewlett-Packard Digital Village Program, TDVnet technicians and managers began working with HPWREN engineers and Hewlett-Packard consultants to build out the backbone. As much as possible, TDVnet project leaders tapped into their circles of friends and associates for local Natives and tribal members to assist with aspects of the build-out. Rantanen described finding the necessary talent in interesting places. For example, casual conversations led TDVnet project leaders to a veteran and helicopter pilot. They contracted the pilot to fly heavy equipment from the roadways up to a mountaintop where TDVnet construction crews were building a tower.

Within a few years, the tribal administration buildings, schools, and libraries had free Internet access. By 2005, the network was robust enough to offset some of the load from the HPWREN relays to the TDVnet backbone. Meanwhile, TDVnet engineers began working with tribal community leaders on designing an intranet archive where members of the nineteen tribes could post photos, news, videos, and knowledge of tribal ways.

When the three-year grant ended, the TDVnet project leaders had a plan in place for maintaining network operations and generating revenues for network improvements. They established an Internet service provider, Southern

California Tribal Technologies, as a tribal enterprise and set up a subscription service for tribal residents and neighbors. They also used remaining grant funds to purchase hardware and software for media labs, a digital print shop, and a professional graphic design studio. They established the print shop, Hi-Rez Digital Solutions, as a tribal enterprise where community members could take basic design classes from the Hi-Rez graphic designer.

TDVnet project leaders also set up a digital recording studio, where community members, and especially youth, could make their own videos, record music, and webcast special events. They set up computing labs, hosting classes ranging from basic computing skills to Cisco network certification courses. Shy youngsters sitting in computing classes learned to use the digital recording studio and were soon editing their own videos and showing friends how to use the studio in the process.

The tribal enterprises began generating profits sufficient to create local jobs and support network enhancements. Community members taking computing classes and using media labs showed increased interest in supporting tribal broadband Internet services. Some tribal members also advanced in job skills training; Southern California Tribal Technologies offered Cisco network certification courses, and leaders began identifying business, technology, and policy solutions for sustaining and expanding operations. They filed for 8(a) certification, setting up SCTT for small-business mentorship, loan, and government contract opportunities. Rantanen partnered with researchers at the University of Illinois at Urbana-Champaign to explore the development of wireless mesh for supplying Internet to tribal residents living in canyons where they could not receive a clear signal from the towers. SCTT engineers wired solar panels and wind turbines into the generators powering the towers, saving on energy costs.

Meanwhile, though the tribal schools and libraries were receiving high-speed Internet through the TDVnet channel, Rantanen discovered that tribal libraries were not qualified to receive e-rate funds—FCC Internet access subsidies for public libraries and schools—because the funding was distributed solely through state governments, not through sovereign Native nations. He began speaking to people in Congress and other political representatives about this and related issues. He gained a seat on the FCC Native Task Force, advising the FCC on adjusting programs so that tribes would get help with acquiring broadband Internet, including the possibility of drafting a tribal priority for broadband spectrum. Indeed, TDVnet was tested and continues to run almost entirely off unlicensed spectrum. This is an extremely important issue that pertains to matters of tribal jurisdiction, the trust relationship between tribes and the federal government, and the often uneasy political relationship between tribes and states, in which states charged with distributing federal funds often overlook, disregard, or are unsure of how to approach tribal neighbors when it comes to resource sharing and funding.

In 2009, when the FCC released a notice of inquiry on how to adjust its broadband grant and loan programs to meet tribal and rural residential needs, TDVnet managers were among the first to point out that the US Department of Agriculture (USDA) Rural Development Broadband Initiatives Program (USDA Broadband Initiatives Program) and National Telecommunications and Information Administration (NTIA) Broadband Technology Opportunities Program criteria were slanted to preclude tribal applicants. While Rantanen was able to demonstrate the feasibility of a proposed TDVnet infrastructural upgrade, SCTT was unfortunately not eligible for an NTIA Broadband Technology Opportunities Program award due to the restrictive program criteria. It did, however, receive funding to support a broadband impact study and digital literacy program through the Zero Divide Foundation, a Bay Area digital inclusion advocacy group. At present, SCTT is increasing demand through sponsoring literacy and learning programs yet still needs major funding to acquire access to spectrum, upgrade hardware at the towers and in labs, and build out the backbone to support greater bandwidth and reach the more remote communities.

Near the end of my visit, on the walk to my car, Rantanen motioned with his hand to show the path of the WiFi signal from the dishes atop the nearest mountain tower to the dishes attached to the tribal administrative buildings. A Hi-Rez Digital Solutions employee had planted sunflowers beneath water trickling from a rooftop swamp cooler. He described how free and affordable access to broadband Internet was allowing cousins who lived on different reservations to connect with one another via Xbox Live, grandmothers to view long-lost photos online, and council members to review a digital archive of past council meetings. ICTs—and especially broadband ICTs—are about helping people connect with one another. For tribal peoples who have been forcibly disconnected from one another for generations by settler-state leaders interested in seizing Native lands and waters, tribally owned broadband infrastructure takes on a value beyond that of simply enabling education, economic development, or cultural revitalization. The combination of business acumen, technical expertise, political savvy, and a few key partnerships helped SCTCA broadband champions grow TDVnet from a demonstration project into a tribal enterprise and create space for ICT innovation and agenda setting in Indian

Country. TDVnet helps many people working at many different locations and through many different positions in the SCTCA tribal community share information and work together toward strengthening the cultural and political sovereignty of the nineteen tribes.

The development of TDVnet is characterized by both the sociotechnical vision of the project leaders and the nature of the partnerships supporting its build-out: it is an example of the social shaping of technical infrastructure. The intertribal and cross-institutional approach introduces a level of complexity that reveals the many jurisdictional matters that tribes must think about when implementing an ICT project of this magnitude.

First, the nineteen tribes that compose the SCTCA represent a diverse range of geopolitical terrain. Some of the tribes are federally recognized. Some are state recognized. Some remain unrecognized by either the US federal government or the state of California, yet as inherently sovereign Native peoples, they bear the rights of cultural sovereignty.

Second, there are differences in the economic capacity of each tribe. Some of the tribes host gaming operations on their reservations. Some tribal governments pay out dividends from gaming operations and other enterprises in the form of per capita payments to individual tribal members. For some tribes, per capita payments can signify tribal households with higher-than-average disposable incomes. Payments can also signify more striking economic differences within a single reservation or neighboring reservations: households with significantly higher incomes can neighbor households below the poverty line. Because of such striking economic variance, I would later find that many tribal ICT champions view their work deploying broadband as a social enterprise—a matter of governance, sovereignty, and self-determination—rather than primarily as a profit-making enterprise.

Third, there are differences of place and nearness to urban centers. In southern California, some of the tribes are located much closer to urban and semi-urban locations that may already receive competitive broadband Internet service rates from regional providers, while others are in more rural or remote locations that lack basic infrastructure. Population counts for each tribe differ, as do community information needs and existing telecommunications infrastructural capacity.

Finally, the nineteen tribes are spread in a checkerboard pattern across the southern California region, meaning that the entirety of the southern California Native homeland is intersected by Bureau of Land Management land, county land, private property, and land that belongs to the state of California.

In terms of deployment, this signifies the drafting of many memorandums of agreement and many right-of-way permits and the consistent promulgation of the vision of affordable and reliable Internet for tribal residents and neighbors.

The leaders of the Southern California Tribal Chairmen's Association agreed to work across these differences when they identified a broadband network backbone as a meaningful long-term intertribal investment. While each tribe may utilize the services of Southern California Tribal Technologies in different ways, the TDVnet project leaders nevertheless set up the backbone to serve all communities regardless of the abovementioned differences. The design and build-out of TDVnet capture some of the best qualities of a community-based network. The design is based on a common vision—connecting tribal peoples for cultural sovereignty and economic development—with the build-out occurring through an iterative series of partnerships, needs assessments, network improvements, and outreach efforts. Figure 4.1 illustrates the SCTCA strategy for acquiring broadband Internet for its constituent tribes. It is based on an intertribal, inter-institutional, collaborative community development approach in which cultural sovereignty and economic development are the impetus, with partnerships, profits from tribal enterprise, and network innovations providing the means of connecting the tribes.

Rantanen's description of the build-out of TDVnet helps us better understand how broadband infrastructures and services designed specifically for tribal communities are conducive to tribal peoples' abilities to connect and create online. Moreover, it shows how the partnerships orchestrated to build these critical infrastructures ground productive relationships between tribal leaders, industry partners, and university researchers.

After speaking with Rantanen and learning about what it took to build out TDVnet, I began to think more critically about the steps tribes have to take if they are to acquire their own broadband Internet infrastructure. I began searching for any kind of documentation revealing the factors shaping the ability of tribal ICT champions to build broadband networks across reservation lands. I found evidence of these factors on the websites of tribal telecommunications and Internet service providers. I found evidence in tribal telecommunications policy papers, those published both on tribal sites and through organizations such as Native Public Media, the FCC Office of Native Affairs and Policy, and the National Congress of American Indians (NCAI). I found evidence in proposed tribal broadband infrastructural deployment plans prepared for the USDA Broadband Initiatives Program and NTIA Broadband Technology Opportunities Program application cycles.

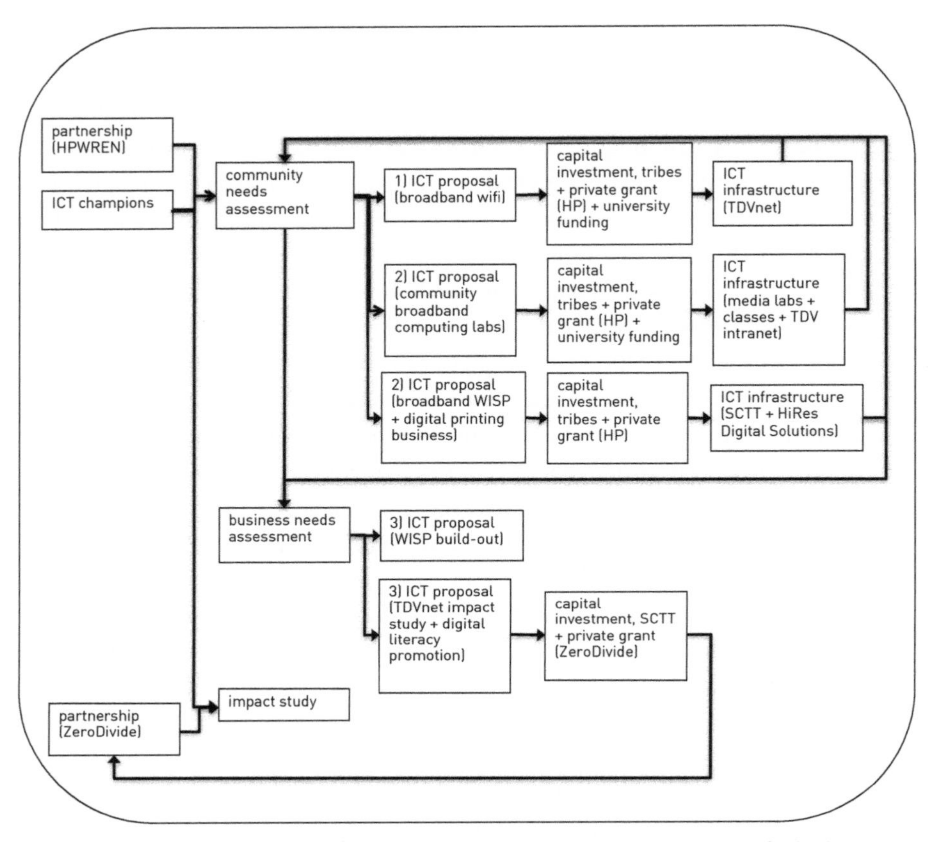

FIGURE 4.1. The Southern California Tribal Chairmen's Association strategy for building a network backbone sufficiently robust to serve the nineteen affiliated tribes consisted of several recursive phases. Local ICT champions partnered with researchers at the High Performance Wireless Research and Education Network at the University of California, San Diego, to build (1) a WiFi network with university funding, grant funding, and a tribal capital investment; (2) a computing lab; (3) a wireless Internet service provider and digital printing business; and (4) a digital literacy project, which led to a new partnership with ZeroDivide.

After a while, I began to get a sense for the conditions shaping tribal broadband infrastructural deployment and selected three more projects to analyze and compare: Red Spectrum Communications (Coeur d'Alene), the Lakota Network (Cheyenne River Sioux Tribe), and the Navajo Nation Tribal Utility Authority and Regulatory Commission. These cases were appropriate candidates because of the length of time they had been operating, the relative durability of their operations, the comparability of their complexity, and their differing approaches to employing broadband Internet to upholding tribal sovereignty.

Moreover, the descriptions of these projects contained clear messages about building ICTs that would support tribal sovereignty as it is imagined in the specific tribal communities. I had read enough studies about technological projects in Indian Country in which researchers obscured tribal sovereignty as a matter of "culture" either by failing to address legal sovereignty at all or by diminishing sovereignty as one of a series of values incompatible with the ideology of techno-scientific progress. Another trend in the literature was the application of methodologies to Indigenous experiences of infrastructure that decentered the history of colonialism from an Indigenous experience. For example, an earlier Marxist analysis of TDVnet produced a compelling analysis but fell short of recognizing tribal goals that move beyond the constraints of nationalist economic orders. This is not to say that the findings of a Marxist analysis do not apply—indeed problems of surplus labor in sociotechnical networks are of central concern in a global networked order. Rather, when it comes to theorizing the impacts of ICTs in Native and Indigenous communities, we also have to acknowledge the significance of digital connectivity for people who were oppressed for generations through an intentional colonial imposition of containment and forced disconnection. In this sense, tribal youth connecting with one another through Xbox Live is quite meaningful. It feels hopeful. It feels possible to imagine that one day those Kumeyaay kids playing Xbox Live might design and develop their own immersive Indigenous game, with all the savvy of tribal youth accustomed to both the strength of their tribal philosophies and the resilience forged in the harsh realities of an economic order built on the nightmare of Manifest Destiny. Foregrounding the early build-out of TDVnet against the history of southern California colonialism allows us to see a moment in which digital connectivity is a privilege that people in tribal communities had not yet taken for granted but which they have begun to breathe into with great intention for a hopeful future.

PART 2. RED SPECTRUM COMMUNICATIONS: ACCESS FOR POLITICAL AND CULTURAL SOVEREIGNTY

In 2011, at the Telecommunications Forum at the Sixty-Eighth National Congress of American Indians in Portland, Oregon, Valerie Fast Horse, information technology director for the Coeur d'Alene Tribe, spoke about some of the issues tribes face when establishing Internet service providers. In an earlier interview published in an Idaho newspaper, she described how her service setting up communication networks for the Army and Army Reserves prepared her to think about potential uses of ICTs in her own reservation community. In the late 1990s, Fast Horse was posted at Dhahran, Saudi Arabia, working as a communications specialist with the US Army. While on active duty, she asked herself, What would happen in the tribe if I were to bring these technologies to the reservation? Hearing Fast Horse speak reminded me of how Rantanen had contracted a military veteran with piloting skills to load and lift heavy equipment from a valley floor to a mountaintop network-backbone base station via helicopter.

Worlds away from Saudi Arabia, Coeur d'Alene is a 525-square-mile expanse sloping between northern Idaho farmland and the eastern Rockies. At the center of the homeland is Lake Coeur d'Alene, a key body of water within a greater watershed, which includes the Coeur d'Alene River and Lake Coeur d'Alene. In 1991, the Coeur d'Alene Tribe filed a lawsuit against mining companies and the Union Pacific Railroad for dumping a century's worth of smelting and mining waste into the Coeur d'Alene watershed. The tribe sponsored a detailed scientific investigation. The results qualified the watershed as the second-largest Superfund site within US borders, with an expected cleanup cost of more than $200 million. While Fast Horse was setting up communication networks in Saudi Arabia, back home in Coeur d'Alene, the Union Pacific Railroad and the Hecla Mining Company settled the resulting environmental lawsuit with the US government. The tribe began leading the cleanup effort in partnership with the US Forest Service, the US Fish and Wildlife Service, the Bureau of Land Management, and the US Geological Survey. By the time Fast Horse returned to the reservation to work as the director of the tribe's department of information technology, the US Supreme Court had recognized that the lower Coeur d'Alene watershed belonged, and had always belonged, to the Coeur d'Alene people. It was a hard-won battle for the Coeur d'Alene people, and it is ongoing, as the tribe leads the cleanup effort.

Fast Horse entered the job bearing a strong message of cultural sovereignty. She understood the power of ICTs for Native peoples, not just as a means of facilitating tribal administrative work practices, but also as a way of sharing the ideas, art, and political commentary that are integral to expressions of cultural sovereignty. With the support of the tribe, Fast Horse created Rezkast, a site where Native people can share videos about matters of interest in Indian Country.

I had heard about Rezkast from tribal librarians, people whose entire professional practice is built around the goals of encouraging alphabetic and digital literacy, not to mention reading for pleasure, in Indian Country. In the first decade of the 2000s, while YouTube was taking off in mainstream society, Rezkast was filling a gap in Indian Country, providing a technical platform for sharing news, language lessons, sports updates, history lessons, and music of interest. Moreover, the prime method of sharing updates through Rezkast is by uploading video and audio files. Native deejays began sharing interviews with Native educators and activists on issues in Indian Country, as well as song files, at the site. Videos of amazing rez ball moments made the archive. So did powwow videos and interviews with elders on different social topics. In one video, Coeur d'Alene elder Noel Campbell speaks about technology and the fear of cameras that many Native people have and explains that he no longer fears cameras, as younger generations are beginning to use computing technology to fight for their rights as Indian people.

The launch and ongoing success of Rezkast helped Fast Horse and tribal IT specialist Tom Jones demonstrate both the capacity of the Coeur d'Alene IT department and the potential for ICT innovation in Indian Country. Recognizing the need for affordable Internet to support tribal household use of technologies like Rezkast, Fast Horse and Jones combined their technical, business, and political acumen to propose a wireless Internet service provider as a tribal enterprise. After conducting a community assessment and demonstrating feasibility, Fast Horse and Jones obtained funding from the tribe and from the 2002 USDA Rural Utility Services Community Connect grant and loan program. They incorporated Red Spectrum Communications as the Coeur d'Alene wireless Internet service provider, offering free or low-cost wireless broadband Internet to community anchor institutions and homes on the reservation and in neighboring rural Idaho and Washington.

Fast Horse and Jones began advising the FCC and other tribal groups on how to think about implementing Internet services for tribal homes and anchor institutions. In light of the ongoing work with the US Geological Survey and other partners, Fast Horse and Jones also expanded Coeur d'Alene IT services to include a geographic information system for surveying and managing tribal lands and waters. In 2011, Boise State University honored Fast Horse as a part of its Women Making History program. It is important to note that in conference presentations, interviews with reporters, and other public-speaking moments, Fast Horse framed the idea of tribal command of broadband Internet infrastructure in terms of addressing specific political, social, and environmental exigencies around Coeur d'Alene, an orientation distinct from mainstream depictions of broadband Internet as a requisite precursor to an American consumer's technological utopia.

During the first five years, after seeing increasing demand for faster broadband, Fast Horse and Jones were already planning an infrastructural upgrade to Red Spectrum operations. After a second community assessment, they developed a plan to deploy fiber-optic cable using the fiber-to-the-home method. This move would not only boost upload and download speeds but also expand the range of service plans offered through Red Spectrum. Much like before, they were able to demonstrate demand, capacity, feasibility, and success with previous Internet service provision to both the tribal council and the 2009 USDA American Recovery and Reinvestment Act grant and loan program for strengthening rural infrastructure toward economic development. At present, Red Spectrum is in the process of laying 275 miles of terrestrial fiber-optic cable that will supply affordable broadband Internet to 3,500 households within the Coeur d'Alene reservation and neighboring communities.

Learning about how Red Spectrum Communications came to be showed me that there are at least a few strategies that tribal ICT champions must employ before establishing broadband Internet infrastructure and services. They must assess community demand. They must stage a pilot project that can demonstrate ICT skill, community impact, innovative capacity, and project completion. For Red Spectrum, Rezkast was the pilot. The outcomes of the pilot project can lead to the development of a proposal for more robust broadband infrastructure. That proposal must take into account available sources of funding, including tribal investments or federal grant and loan awards. Each time a phase of infrastructural build-out is finished, the project leads assess the outcomes, synthesizing that knowledge until it is time to plan and propose the next network enhancement. Figure 4.2 depicts the Coeur d'Alene strategy for acquiring broadband Internet access.

The exercise of cultural sovereignty is of prime importance to the Coeur d'Alene Tribe and guides administrative operations and investments in enterprise. Coeur d'Alene recognizes the sovereignty of tribes as inherent. Unlike tribes whom the US government forcibly removed from their homelands, the people of the Coeur d'Alene Tribe lived within the lands now known as Idaho and the United States long before the establishment of either. At present the Coeur d'Alene Tribe's commitment to protecting the homeland is enforced through the exercise of legal and political sovereignty, but especially by investing in projects that, like Rezkast, strengthen Coeur d'Alene cultural sovereignty. While Red Spectrum Communications has been funded by the USDA as a matter of rural economic development through infrastructural improvement, Rezkast, the tribal global information system, and Red Spectrum were sponsored and

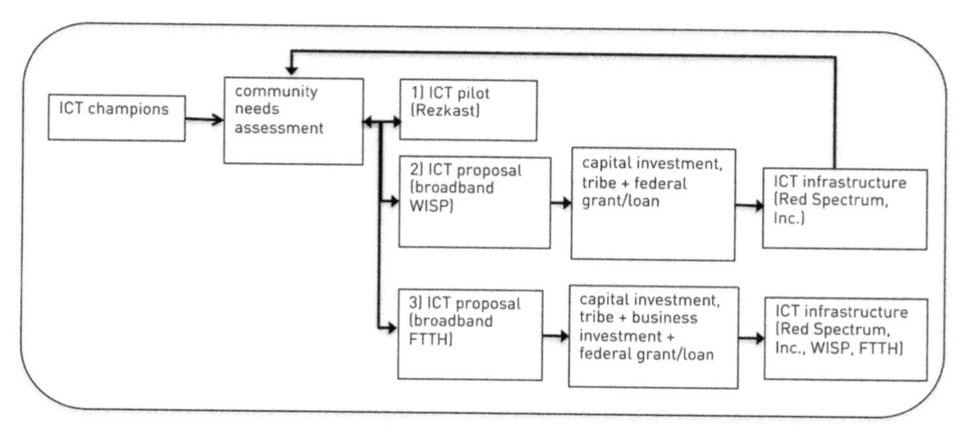

FIGURE 4.2. The ICT champions at Red Spectrum Communications worked through their capacity within the Coeur d'Alene tribal government to conduct a community needs assessment and pilot Rezkast. The success of the pilot gave them leverage to propose a tribal wireless Internet service provider and acquire a federal loan as well as a tribal capital investment. The success of Red Spectrum Communications led to their fiber-to-the-home project.

funded by the tribe as an investment in the Coeur d'Alene people, who need to communicate with one another about matters affecting their ability to defend their cultural integrity and the health of their homelands and waters. The Coeur d'Alene Tribe's approach to broadband Internet deployment is based in tribal governance goals and funded in part by tribal business revenues, including those of Red Spectrum and other enterprises. The technical approach is based on leapfrogging and infrastructural build-out over time in step with community demand and tribal governance goals. Red Spectrum's focus is on improving access to broadband so that Native peoples within and beyond the Coeur d'Alene homelands can support tribal sovereignty and cultural revitalization.

Comparatively, tracing the build-out of Southern California Tribal Technologies' TDVnet and the Red Spectrum Communications network backbone teaches us about the iterative process of building out networks of this scale. Cycles of visioning, needs assessment, deployment, and modification, proposals for enhancement and support, and processes of investment lead to many opportunities for both strengthening the vision for use of these technical infrastructures and stabilizing their value within the values frameworks already in place within the communities of use.

Tribal broadband networks are distinguished from other community-based broadband networks by values around duty to tribal homelands and place-based ways of knowing. Native peoples continue to serve in higher than average numbers in various branches of the US military. Learning about ICTs in Indian Country revealed an aspect of the contributions of veterans in tribal communities. Many veterans return from active duty with a range of digital communications and leadership skills, the desire to contribute to their tribal communities, and eligibility for veterans loans through the US Small Business Administration. While I observed multiple cases of veterans supporting ICT projects in Indian Country, in the case of Red Spectrum Communications, Valerie Fast Horse dedicated her particular skill set, respect for homelands, and lived experience to building out an entire network for her tribe.

Understanding the Red Spectrum Communications build-out teaches us about what happens in stages through the build-out process. Once the broadband network is woven into the tribal homeland, the reservation gains a digital overlay and becomes a topography for knowledge sharing and memory work through both digital and face-to-face modes of communication. With each change made to the network itself, as well as with the introduction of newer systems and devices—faster phones, flashier social media interfaces, global information system tools, online grants and procurement systems—there is an attendant calibration within the communities of use: a time of practice, modification, assessment, and acceptance (or not) of specific devices and interfaces. The question is thus not whether tribes will accept this new technology but rather whether tribal communities will incorporate these systems into existing work and life practices, in accordance with existing ways of knowing.

PART 3. MANY VOICES, MANY SOLUTIONS AT THE 2012 TRIBAL TELECOM AND TECHNOLOGY SUMMIT

Two years into this research, I quickly gathered that there was a loose affiliation of like-minded individuals working on tribal Internet and tribal telecommunications issues in Indian Country. The same names, federal offices, and associations kept popping up not only in descriptions of their work but also in policy papers. As an information scientist, I had the sense that this was perhaps evidence of networks of individuals practicing an effective form of agenda setting, tapping into their own social networks within their communities in order to connect with individuals in federal offices. However, the 2012 Tribal Telecom and Technology Summit showed me another layer of complexity in the political landscape around telecommunications and technology in Indian Country.

I would come to understand that this layer of complexity had to do with (1) the diversity of tribes and the accompanying varying modes of decision making, (2) choices about how to conscript various kinds of networked computing systems into specific tribal governmental processes, and (3) the nature of investing in perpetually advancing digital technological systems. I found that those working at the intersection of tribal policy and telecommunications deployment are able to gain perspective on decisions affecting their local communities by sharing experiences and ideas in goal-specific intertribal forums. The first realization: when tribal leaders begin thinking of incorporating streaming radio, high-speed computing labs, online voting, social media, or other such networked computing technologies into their tribal community or government services, they start to think about the availability of the Internet for tribal members. Thinking about that brings up questions on the cost of devices, areas of Internet availability, digital literacy, privacy and security, technical skills development, and more. It becomes apparent that the introduction of networked computing systems into jurisdictionally bounded spaces like reservations opens up a host of questions about Internet practices, policies at many governmental levels, economic development, education, and more. Moreover, conversations about these topics need to occur at the same time across various levels of leadership within and around tribal communities. A kind of complex communication occurs, and the team who posed the original idea functions as the spider testing various threads in a web of related conversations.

I first experienced this somewhat overwhelming stream of conversations in 2012, at the first Tribal Telecom and Technology Summit in Phoenix. I went seeking documentation of any US tribal broadband operation reaching the K-Net level of specificity. The conference was organized by just a few groups: a tribal telecom taxation law firm, Native Public Media, and Gila River Telecommunications, which was at the time one of the handful of tribal Internet service providers in the United States. The conference was hosted at the Wild Horse Pass Resort and Casino on the Gila River reservation southwest of Phoenix. While there, I met many attendees who were hoping to find out more about either acquiring basic telecommunications services for their reservation communities or overcoming obstacles to setting up tribal Internet service providers.

A representative from Havasupai was there to learn about how her tribe could acquire access for communities located at the base of the Grand Canyon and high up through rocky terrain. I also met two men who described how a private telecommunications company established a monopoly across their desert reservation in California years ago and committed the tribe to a noncompetition agreement in exchange for much-needed telephone service. Later, when the tribe built a casino and resort, leaders realized soon enough that members were being significantly overcharged for Internet and phone service. By that time, the tribe had the capital, technical personnel, and know-how to set up its own Internet service provider, and at much more affordable rates than those of the predatory company. Unfortunately, the noncompetition agreement was still standing and, with it, the threat of costly legal battles.

Summit attendees included managers of Internet service providers who had successfully navigated the FCC spectrum licensing process, lawyers and accountants whose entire work focused on detangling tribal telecommunications taxation issues, community leaders whose elders expressed concern about the distractions ICTs might introduce among tribal youth, lawyers thinking about how—because of the nature of online information sharing—tribal telecommunications and Internet service providers interrelate with tribal intellectual property issues, and entrepreneurs building data centers as part of tribal economic development portfolios. There were even designers of wireless towers who specialized in constructing towers that merge aesthetically into the local built environment: towers shaped like trees, designed as public art, or in the color palette of neighboring buildings.

Many of these individuals noted the lack of reliable public data about telecommunications and Internet services in Indian Country. Many also remarked on a critical discrepancy in federal subsidy programs. On the one hand, the federal government provides subsidies for building out basic landline telephone and 911 services to low-income and rural residents. Many residents of Indian Country are low-income and live in rural and remote locations, so this would seem to be a helpful option that tribes would want to pursue. On the other hand, the federal government also provides grants and loans for broadband Internet infrastructural build-out in rural and remote locations. The National Broadband Plan spells out a strategy for providing a majority of critical services—Internet, energy and electricity, economic development, citizen participation, law enforcement, education—across broadband Internet modalities such as WiFi, satellite, and fiber-optic cable, and less so through basic landline telephone infrastructure.

This positions cash-strapped tribes between a rock and a hard place. Acquiring basic phone service subsidies requires showing need and lack of telecommunications infrastructure, yet acquiring broadband Internet infrastructure subsidies requires showing demand and feasibility. This uncomfortable space—somewhere between total poverty and total possibility—is unfortunately a familiar space for tribal leaders seeking grants and loans. It is not unrelated to the colonial formation of an "Indian problem": colonial authorities will grant rations if the people can prove death and devastating illness, but as soon as they demonstrate a bit of health, they are constrained by the bureaucratic red tape that maintains the disconnections imposed by the reservation system.

Summit organizer and tribal taxation lawyer Randy Evans later said that the 2012 conference was originally planned to be a small workshop for fewer than forty individuals. But word spread, and registration quickly mushroomed to more than a hundred participants, transforming the Tribal Telecom and Technology Summit from a workshop to a conference. FCC chairman Michael Copps spoke at the first conference, underlining a commitment that former president Bill Clinton had made during a 2000 visit to the Navajo reservation: the federal government has a responsibility to help tribes connect and overcome the digital divide. In 2013, due to a federal sequester, the FCC was unable to send representatives from its recently formed Office of Native Affairs and Policy. Nevertheless, conference registration and attendance at the 2013 summit were nearly triple the 2012 summit registration and attendance. Evans described how telecommunications and Internet service provider taxation topics alone are complex and important, a point underscored by Shana Barehand Green. I recalled that Traci Morris, director of operations for Native Public Media at the time, referred to unexamined vast and thorny tribal broadband policy issues, especially with regard to the sovereign rights of tribes, and how each tribe chooses to enforce and enact those rights given its political geography.

The Tribal Telecom and Technology Summit underscored not only the differing values and experiences shaping tribal decisions to pursue broadband build-out but also the common goal of building these infrastructures as tools in support of sovereignty and self-determination. As the 2012 conference progressed, speakers and participants noted that the call for "best practices" in infrastructural development seemed unrealistic; that every tribe seemed to identify its solutions for acquiring broadband and telecommunications services based on its own unique tribal histories, geographies, terrain, and existing levels of infrastructural access.

Thus in the case of tribal broadband deployment, the idea of place-based ways of knowing refers to not only ecological knowledge and tribal philosophies but also knowledge of the local political, social, economic, and technical terrain. At present there are more than 568 federally recognized tribes within US borders. This does not account for tribes that are state-recognized and those that are unrecognized by federal and state governments yet bear the rights of inherently sovereign Native peoples. In the years to come, each of these communities—tribes and peoples—will apply broadband access solutions according to its technical capacity, unique landscapes, and needs. There should be more than 568 different solutions to acquiring broadband access in Indian Country.

Over the next few years, Tribal Telecom and Technology Summit organizers would embrace this finding, acknowledging that many voices contribute many inspiring and instructive experiences of tribal telecommunications and broadband deployment, from infrastructural concerns to application-layer creativity like Rezkast and streaming radio. They would also cleave firmly to a resolution foregrounding the experiences of presenters working for tribal sovereignty and self-determination and avoiding sales presentations by wireless vendors or how-to presentations by federal authorities far removed from tribal life on reservations. Likewise, as a researcher, attuning oneself to the nature of leadership in Indian Country—a leadership fully aware of US colonial mentalities and what it takes to work through and think beyond colonial expectations—opens up the potential for understanding how to design methodologies and garner insights that reflect Native experiences and perspectives.

Understanding the formation of the Tribal Telecom and Technology Summit as a forum designed by individuals invested in applying ICTs toward tribal sovereignty reveals another layer of social power enacted through sociotechnical infrastructural development. Network backbones, Internet service providers, and the accordant policies and practices represent such complex systems that it does indeed take many people to make them work. In the case of reservation-based systems, all those individuals become conscripted in one way or another into the pursuit of tribal sovereignty and self-determination. Those who obstruct tribal build-out of Internet and telecommunications infrastructure stimulate the organization of tribal advocacy groups and their allies, leading to the eventual formation of expert telecommunications policy advocates working across multiple reservation communities, multiple levels of government, and educational institutions and in Washington, D.C.

Through these conferences, one can see the opening of an industry in Indian Country, including a fair share of economic uncertainty, entrepreneurial acumen, big dreams, and technical know-how. On occasion, conference organizers and presenters would ask participants questions like, "Is this really happening? Do Indians really have a seat at the table for making these decisions?" Inevitably, the director of a successful tribal telecom or Internet service provider—perhaps Ruben Hernandez from Fort Mojave Telecommunications, Godfrey Enjady from Mescalero Apache Telecom, John Badal from Sacred Wind Communications, or Danae Wilson from the Nez Perce Information Technology

Department—would stand and affirm the reality of the digital overlay, its profitability, its potential for stable economic development, and the need for continual training and technical skill development.

PART 4. LAKOTA NETWORK: TELECOMMUNICATIONS FOR ECONOMIC SELF-DETERMINATION

While TDVnet and Red Spectrum Communications were designed around strong values of cultural sovereignty, and TDVnet incorporated elements of economic development, Lakota Network out in Cheyenne River Sioux country is equally compelling because of its stronger emphasis on telecommunications as a matter of economic self-determination. Several speakers and participants at the Tribal Telecom and Technology Summits mentioned that tribal investment in ICTs creates job skills, but the idea that ICTs might provide a means toward tribal economic self-determination is much greater than that. This idea of ICTs implies the capacity for a tribal community to embrace a life world wherein there is sufficient technical skill, demand, and supply of digital goods such that a tribe can gain a corner in a global telecommunications market. The Cheyenne River Sioux Tribe has managed to do just that. In 1958, the tribe established the Cheyenne River Sioux Tribe Telephone Authority (CRSTTA). One of the first 100 percent Native-owned telecoms, it was sponsored by the tribe but operated by a board of directors separate from tribal administration.

The Cheyenne River Sioux Tribe has enforced tribal sovereignty through negotiations with the United States for more than a century, and perhaps most famously through the 1868 Treaty of Fort Laramie. Awareness of representation and the power of infrastructure pervades the language of the treaty. Article XI requires that tribal signatories

> relinquish all right to occupy permanently the territory outside their reservations as herein defined, but yet reserve the right to hunt on any lands north of North Platte, and on the Republican Fork of the Smoky Hill River, so long as the buffalo may range thereon in such numbers as to justify the chase.

Article XI outlines seven requirements to which the "said Indians further expressly agree." The sixth requires that

> they withdraw all pretence of opposition to the construction of the railroad now being built along the Platte River and westward to the Pacific ocean,

> and they will not in future object to the construction of railroads, wagon roads, mail stations, or other works of utility or necessity, which may be ordered or permitted by the laws of the United States. But should such roads or other works be constructed on the lands of their reservation, the government will pay the tribe whatever amount of damage may be assessed by three disinterested commissioners to be appointed by the President for that purpose, one of the said commissioners to be a chief or headman of the tribe.

Since that time, the four bands of the Peoples of the Plains—the Mnikoju, Owohe Nupa, Itazipa Cola, and Siha Sapa—have negotiated with US and Canadian federal authorities and civil authorities through the states of South Dakota, North Dakota, Montana, Wyoming, Nebraska, Iowa, and Minnesota on matters affecting the right to access sacred sites, protect the homelands, repatriate artifacts, exercise religious freedom, and provide for the just and lawful treatment not only of the people of the Cheyenne River Sioux Tribe but also of the Native peoples of Turtle Island.

During the late 1960s and early 1970s, much organizing around the American Indian Movement (AIM) happened through the Nakota-Dakota-Lakota homeland. Keenly aware of the impacts of media and telecommunications, AIM organizers mobilized Native peoples and allies across Turtle Island through strategic radio and television broadcasts. The generation of Native leaders who are now running major political organizations, such as the National Congress of American Indians, and directing the operations of tribal colleges and other national-level forums were in their teens when the AIM occupation of Alcatraz and the federal government blockade at Pine Ridge were shown on television. Both of these broadcasts irrevocably shaped the nature and ethos of intertribal organizing and activism in Indian Country. The scholarly study of media representation continues to be a significant area of research in American Indian Studies.

The historical understanding of tribal media and ICT infrastructure as a mechanism for the enactment of tribal sovereignty also pervades the establishment of the Cheyenne River Sioux Tribe Telephone Authority. It was the first Native-owned company to utilize loans from the Rural Electrification Administration—one of President Roosevelt's many New Deal programs—to improve services for tribal residents. The establishment of the tribal telephone authority parallels the establishment of the tribal radio station. By the late 1970s, the telephone authority had a strong record of engagement with the Rural Utility Services loan programs. For decades, CRSTTA leadership expanded the telephone authority, investing in the business as a tribal enterprise and training Native employees. A commitment to self-determination through tribal economic development fueled continual investment.

By the late 1990s, as the CRSTTA was establishing itself as an eligible telecommunications carrier with the FCC in accord with the 1996 Communications Act, telecommunications entrepreneur and CRSTTA manager J. D. Williams (Cheyenne River Sioux) was also challenging the South Dakota Public Utilities Commission's attempts to regulate the sale of telephone exchanges on reservation land as an infringement of tribal sovereignty. Around this same time, tribal IT specialist Gregg Bourland (Cheyenne River Sioux) began persuading tribal administrators to set up websites for their departments. While this proposal was met with doubts at first, over time, tribal administrators saw the benefits of Internet access and self-representation in the online environment. Within a few years, Bourland had advanced politically to become a member of the tribal council. He supported the development of a tribal department for the management of information systems. He also began working with Williams to think of ways to start up an Internet service provider for the tribe. After conducting a community assessment and writing a business plan, he and Williams convinced the board of directors to commit funds and allow employee training. Thus the Lakota Network was established as a regional Internet service provider. By the mid-2000s, the CRSTTA began laying miles of fiber-optic cable through the reservation.

Tracing the development of the Lakota Network over time reveals how the current infrastructure is an outcome of tribal leaders' many negotiations with state and federal authorities. Through infrastructure construction, tribal leaders enforced Cheyenne River Sioux rights to self-governance and self-determination as these relate to telecommunications. In that way, tribal leaders positioned federal support of telecommunications infrastructural loans to tribally owned business as part of the federal government's trust responsibility to US tribes.

Within a few years of establishing the Lakota Network as an Internet service provider, Bourland and Williams begin thinking about how to increase revenues to support additional infrastructural build-out and training so that the tribe could create businesses based in a knowledge economy. They drew up a business plan for hosting a credit card company's data management and backup services through the Lakota Network. Unfortunately, negotiations fell through when Bourland and Williams realized there was insufficient technical skill among the reservation workforce. I appreciated learning about this dimension of long-term broadband infrastructural build-out; failures and false starts are also learning opportunities and encourage new perspectives on growth. Without a firm faith in a self-correcting future, it is possible to be overwhelmed by a false start. In this case, Bourland and Williams were seeking to create a future in which tribal citizens could flourish through knowledge work, becoming digitally engaged employees and businesspeople from their reservation community.

Thus Bourland and Williams learned from the failure, conducted another community assessment, and convinced the tribal government to invest in a community computing and training center that would focus on increasing the technical skill set of tribal members. While tribal members underwent certified training programs, Bourland and Williams drafted another business plan to spin off a data entry, document digitization, and digital records management company, Lakota Technologies, Inc. Their plan passed muster with the tribal council, the telephone authority board of directors, and the USDA American Recovery and Reinvestment Act grant and loan program committee. With funding from the tribe and the USDA, Bourland and Williams established Lakota Technologies, Inc. They also acquired prime data digitization contracts from the National Library of Medicine, the US Department of Defense, and other key partners. Williams has since retired, and Mona Thompson (Cheyenne River Sioux) now manages the Cheyenne River Sioux Tribe Telephone Authority. At this point in history, technical improvements to the Lakota Network occur through investments in Lakota Technologies Incorporated.[3] Figure 4.3 depicts the Cheyenne River Sioux Tribe strategy for acquiring broadband Internet access.

Like the Coeur d'Alene Tribe and the Southern California Tribal Chairmen's Association, the Cheyenne River Sioux Tribe established an affordable Internet service provider as a tribal enterprise. However, unlike Red Spectrum Communications and Southern California Tribal Technologies, the Cheyenne River Sioux Tribe Telephone Authority is primarily a for-profit business-driven model focused on investing in tribal ventures in knowledge work. The strategy for building out the Lakota Network and developing a robust broadband infrastructure on the reservation follows cycles of business planning, community readiness assessment, and opportunities for increasing the tribe's return on investment. The long-term plan is to diversify the tribal business portfolio.

By developing lucrative business partnerships, acquiring government contracts, and capitalizing on profits in order to attract government grants,

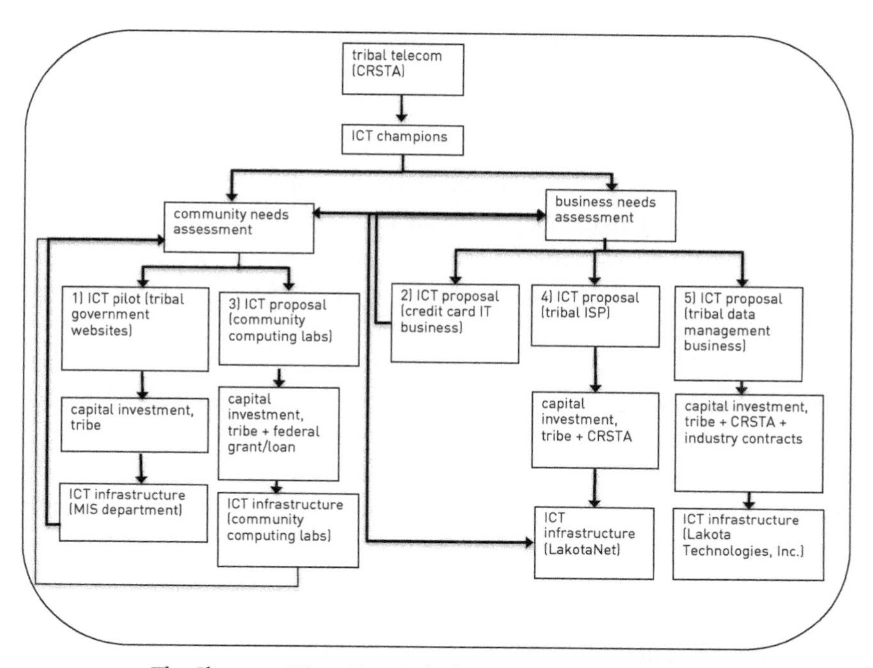

FIGURE 4.3. The Cheyenne River Sioux Telephone Authority engaged in business planning in addition to community needs assessment as leaders developed a strategy for acquiring broadband Internet. While their focus on community needs led to the development of their tribal government websites, management of information systems department, and community computing labs, their business planning led to a credit card billing business, which gave them the opportunity to grow the skill set for a tribal Internet service provider and, eventually, Lakota Technologies, Inc.

the CRSTTA is creating knowledge work opportunities for tribal members. The Cheyenne River Sioux Tribe's approach to leveraging telecommunications so as to bring in profits and create jobs on the reservation is a distinctive expression of economic self-determination. While self-determination refers to the right of tribes to design and implement their own social services programs for their people, economic self-determination refers to the right of tribes to support tribal enterprises that best meet community needs. The Cheyenne River Sioux Tribe's investment in broadband Internet services, geared toward economic self-determination, relates to the inherent sovereign right of the four bands of the Peoples of the Plains to live and work within their homelands.

This case is also helpful for revealing another quality about the stability of large-scale community-based ICT infrastructures. In many ways, the effectiveness of Lakota Technologies, Inc., is built on the effectiveness of the Lakota

Network, which was in turn built on lessons learned through both stable and riskier innovations and investments through the telephone authority. Similarly, the effectiveness of Red Spectrum Communications at Coeur d'Alene was built on lessons learned through the deployment of Rezkast. The effectiveness of the TDVnet built out of Pala was based on lessons learned through the deployment of HPWREN. This demonstrates that if a tribe already has some effective digital system or infrastructure in place, the path its ICT project team chooses for broadband deployment will be shaped in part by its experiences with the existing digital system and technical infrastructure.

In ICT circles, this reflects the sensation of "always being in beta," the concept that every project is a pilot until it becomes reliable and useful enough to serve as a staging ground—on the basis of either technical infrastructure or the accumulation of sociotechnical expertise—for another project. John Law and Bruno Latour refer to the exercises of power inherent in the accretional and cumulative design of sociotechnical systems, when the system begins to generate its own normative practices and rules of logic and customary order.[4] This is the calibration to which Richard Coyne refers, pilot technology projects—in these cases, accessible interfaces like tribal websites, landline phone services, and Rezkast—become communicative instruments allowing community members to become attuned to one another in new ways and also to improve the interfaces to meet community needs through regular work processes of modification and innovation.[5] Tribes who support the incubation of home-grown ICTs are thus also creating environments that allow for this community-level calibration. Here we see the appropriateness of a community informatics approach to system design, in which tribal communities that command the means for designing, deploying, and using sociotechnical systems also appear to have command over the ordinary, daily practices of calibration that integrate these systems into the local ecology of work practices.

Here we also see the need for Indigenous scholars to understand the ways global economic patterns shape decisions about tribal ICT investments and the access that Indigenous activist-advocates and students have to digital communications infrastructure. In one way, creating jobs on reservations that develop digital literacy is a necessary idea. Tribal residents can enjoy good jobs without having to leave the reservation, and if they do choose to go elsewhere for work, they will have a skill set that qualifies them for fair-paying jobs in the knowledge economy. This represents a goal for tribal leaders who are concerned with overwhelmingly high unemployment rates on reservation lands. From another perspective, however, Indigenous scholars worry about the reach and influence of neoliberal markets in Native communities, particularly through prosumption, the willingness of social media users to contribute free labor through their social media use while media companies profit. ICTs certainly represent a world of possibility for the unjust neoliberal circulation of labor, bodies, information, and goods in world trade circuits. Many decolonization goals are fixed on the eradication of unjust neocolonial trade and labor practices and their replacement with alternative pathways to economic development, including local, environmentally sustainable, communal approaches. These kinds of questions make the study of ICTs critical in Indian Country. The leadership at the Cheyenne River Sioux Tribe Telephone Authority has undoubtedly supported moves for self-determination around telecommunications in Indian Country. J. D. Williams has testified many times before Congress and has made sure that the state of South Dakota does not profit unduly through exploitation of the rules of sovereignty. We have to ask ourselves how tribal command of ICT infrastructure and enterprises shapes tribal social, political, and economic power sharing under conditions of neoliberal colonialism. Having a greater vision of tribal ICTs as part of global economic circuits of trade helps us understand the span of these impacts as we weave broadband infrastructures into tribal landscapes. We should not shrink from these questions but, rather, embrace them as we allow the diffusion of digital devices through our homelands.

PART 5. NAVAJO NATION: REGULATING TO PROMOTE COMPETITION

Without a doubt, when tribes invest in large-scale infrastructures for public goods—waterways, electric power, transportation infrastructure, and communications infrastructure—they are also investing in creating a space for themselves in global markets. The build-out of the Lakota Network showed me the importance of tribal command of these kinds of infrastructures, from ownership, to citizen training and employability, to decisions on how to spin off associated tribal enterprises. Indeed, command of the infrastructure and ownership of the associated businesses are what allowed the Cheyenne River Sioux Tribe Telephone Authority to acquire contracts from the National Library of Medicine and other government clients. This highlights an important aspect of tribal broadband networks: though the design and development of tribal broadband infrastructures are community-based, the resulting Internet service provider is very much a business venture. Intertwined in political, cultural, and legal terms, tribal sovereignty and self-determination are fundamentally about a people's right to provide for themselves in ways that best meet their needs. Indeed, designing a project based on the sovereign rights of tribes and values of self-determination represents one antidote to the so-called Indian problem.

But the "Indian problem" is multidimensional. It is the metaphor that elite classes in a white supremacist state utilize to describe the black box of Indian Country. It is a rhetorical trope that, in one phrase, capitalizes on societal ignorance about the specificity and diversity of many tribal nations, histories, philosophies, and landscapes. When tribal leaders work together to challenge this trope, pointing to dominant society for the cause of their exigency and then identifying home-grown solutions that work better than the self-serving solutions offered by members of the global elite, radical transformations are possible. New solutions emerge that center the needs and potential of tribal ways of life.

With regard to ICTs in Indian Country, one of the most well-publicized cases of tribal lack of Internet access began with a story about Myra Jodie, a teenager from the Navajo Nation who won a computer she could not use. That a teenager could not use a new computer in her home stunned an American public fascinated with its own burgeoning global digital reach. Media attention resulted in the Gates Foundation Native American Access to Technology Program, which benefited my own relatives on the Pascua Yaqui reservation and which I, as a librarian, had been following for years.

In the winter of 2012, I attended a session of the Tribal Telecom and Technology conference hosted by Navajo Nation Telecommunications Regulatory Commissioner Brian Tagaban. Tagaban had been a Cisco network administrator before he returned to Navajo Nation to encourage the tribal broadband build-out strategy. The Navajo Nation Tribal Utility Authority and the Navajo Nation Telecommunications Regulatory Commission are important cases for consideration because of the unique approach to tribal regulatory command of broadband infrastructure and the potential for broadband innovation and enterprise within Navajo Nation. Fitting the story of the Navajo Nation Telecommunications Regulatory Commission alongside the story of Myra Jodie's big win teaches us something about Native uses of ICTs in the popular imagination.

The story begins in 2000, when fourteen-year-old Myra Jodie used a computer at her school on the Navajo reservation in Arizona to enter a contest for an iMac. Actor Jeff Goldblum was advertising the iMac's ease of use: plug and play, hard drive and monitor in one, a good fit for every home with a phone line. But Myra's family did not have a phone line. The San Jose–based contest sponsors traced her home address and from there contacted her school. However, once the iMac was delivered, there was still the problem of the phone line. What use was this computer in a place where electricity was at a premium, and where Internet access was limited to a few machines at the school?

Myra Jodie's contest win became iconic for digital divide advocates. Former president Bill Clinton recognized the incredible divergence in access highlighted by this case. In April 2000, he became the first US president to visit a reservation, the Navajo Nation, where he specifically addressed issues of the digital divide in Indian Country. While Apple was airing ads showing how easy it was to plug and play on a bright-hued Mac, here was a story of a young person who owned a Mac but could not easily get access to a basic phone line. In response, community development program officers at the Bill and Melinda Gates Foundation, the corporate giving branch of Microsoft, implemented the Native American Access to Technology Program (NAATP), through which the foundation would provide Microsoft hardware, software, training, and funding for setting up local area networks for US tribes.

However, the rural tribes of the Four Corners region—where the corners of New Mexico, Arizona, Utah, and Colorado meet—posed unimagined problems to NAATP managers. Local area networks were either impossible to set up or unaffordable in locations without cables for landline phone service or, in some cases, electricity. While NAATP technicians synced expensive satellite hookups and hosted community training sessions, tribal personnel wondered to what end they would make use of these costly machines when their analog work practices were already well suited to tribal daily life. NAATP officers noted in a mid-point project evaluation that every tribe they encountered spoke of the incipient "smoldering conflict" in border towns and schools that was in part a factor in community concerns about adopting potentially exploitative ICTs. By the time funding for satellite Internet access ended, many of the tribal communities had resorted to managing the computers in a somewhat limited fashion. In a final report, somewhat overwhelmed by the complexity of their effort, NAATP officers concluded, "We know that we still have more to learn, so we are concentrating on the work ahead of us before we decide whether to expand our scope. We do know that the Gates Foundation and those of us who work for it have benefitted greatly from this program. Our perspective and understanding has broadened and deepened. We believe the best way we can honor the inspiring efforts of the tribal libraries and librarians we have met is to share our story and to support their work." Unfortunately, in Indian Country, this kind of refrain is common. It as if "our problems" are too complex for a quick fix, such that obstacles are more like endemic conditions than solvable problems. I detected elements of an "Indian problem," in which justice-oriented members of the privileged class either ignore or cannot perceive or make sense of the impacts of colonialism in their diagnosis of why and how charitable projects fail. These kinds of diagnoses often pin the problem on the tribal peoples themselves, hence the regenerative violence of the "Indian problem" and projects designed to cure it.

However, while the Native American Access to Technology Program was attracting mainstream media publicity, in other parts of the Navajo Nation, tribal ICT champions were drafting a proposal to acquire technology contracts from the US Department of Defense and partnerships with HPWREN, the TerraGrid project, and the US Department of Energy. These contracts would lead to investment, including training, equipment, and lab space, in Navajo Technical College, a two-year institution focused on increasing the technical capacity and employability of Navajo Nation residents. Learning in part from challenges encountered through the NAATP, and also in line with Navajo Nation goals to support educational and economic opportunities for tribal members, they wrote the Internet to Hogan plan, a ten-year initiative to install the largest wireless mesh network on a reservation through chapter houses and other key anchor institutions. Since well before the famed achievements of the Code Talkers, the people of the Navajo Nation had been working on what would now be considered cyber-security and protection of the land through intelligence gathering and policy making. Sticking close to the long-term plan to embed Navajo Nation with a durable ICT infrastructure and support the technical advancement of the Diné people, the Navajo Nation Tribal Utility Authority has moved on to another phase of broadband deployment, a fiber-to-the-home project funded in part by a USDA Broadband Initiatives Program grant and loan award. The nation also split up ownership and regulatory oversight of acquiring broadband Internet access for tribal peoples by establishing a separate Navajo Nation Telecommunications Regulatory Commission.

Soft-spoken and knowledgeable, Tagaban developed his business and technical acumen working at Cisco as a network administrator. He returned to Navajo Nation to serve on the regulatory commission, guiding the nation on matters related to the close connection between technical and policy decisions around broadband network design, deployment, and use. Unlike the broadband deployment strategies of Coeur d'Alene, the Southern California Tribal Chairmen's Association, and the Cheyenne River Sioux, Navajo Nation has not started its own profit-making Internet service provider as a tribal enterprise but rather

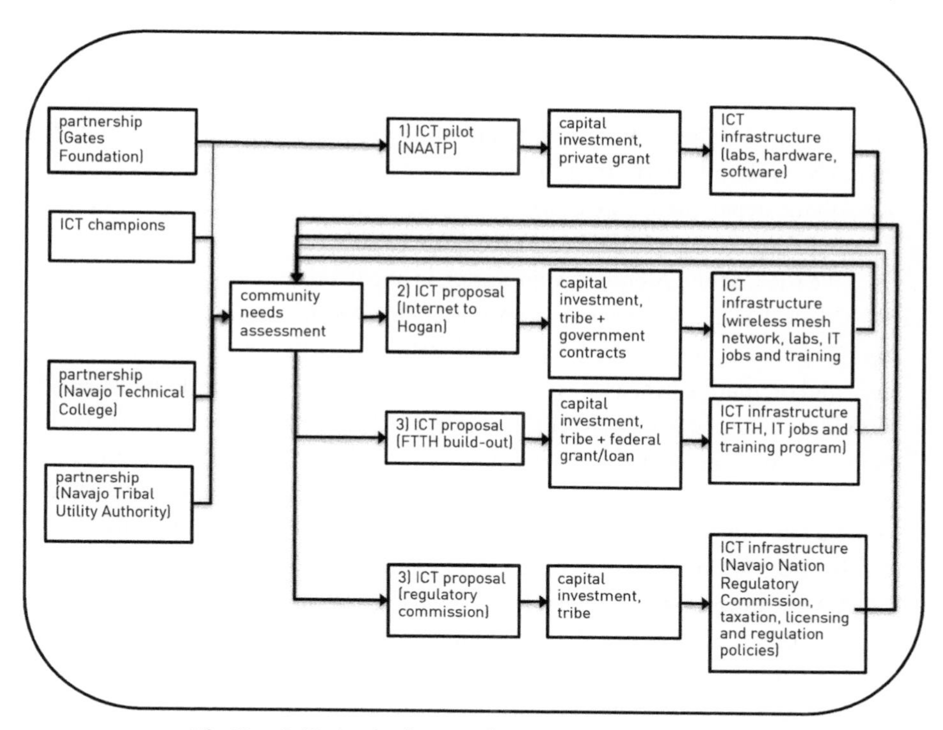

FIGURE 4.4. The Navajo Nation implemented its strategy for acquiring broadband Internet infrastructure through multiple partnerships. ICT champions considered the outcomes of the Gates Foundation Native American Access to Technology Program in addition to community needs assessments, which in turn led to the Internet to the Hogan plan, fiber-to-the-home build-out, and tribal telecommunications regulatory commission.

has chosen to take ownership of the infrastructure and regulation of its use while taxing external Internet service providers that use the reservation's broadband infrastructure.

The goal of this approach is to promote local competition among service providers, many of whom, given access to durable reservation infrastructure, receive federal and state subsidies for improving access across the neighboring states and rural counties of Arizona and New Mexico. The increased competition can lead to lower pricing for residents of the nation and also relieve the tribal utility authority and regulatory commission so they can continue focusing on infrastructural build-out, enhancements to Navajo Technical College, policy work, and job creation and training. At the 2012 Tribal Telecom and Technology Summit, Tagaban indicated the complexities of having to both train and certify IT specialists as network database administrators, which is technically a whole new ball game, while keeping track of broadband policy changes at the national level as they interface with changes at the tribal level. He split his time between training, updating chapter house leaders and key tribal broadband partners, and drafting and tracking policies to keep the broadband deployment efforts in motion. Figure 4.4 depicts the Navajo Nation's strategy for acquiring broadband access.

Like the Coeur d'Alene, Navajo Nation is one tribe and one people. Unlike the Coeur d'Alene, however, Navajo Nation decided not to set up an Internet service provider as a tribal enterprise but rather to invest in durable infrastructure and regulation in order to incentivize competition among regional service providers. Like the Southern California Tribal Chairmen's Association, Navajo Nation acquires access to infrastructure, hardware, and software by leveraging partnerships through educational institutions. However, although the SCTCA and Navajo Nation both must deal with agreements across jurisdictional lines, the boundaries are distinct: intertribal and inter-institutional in the case of SCTCA and intratribal and interstate in the case of Navajo Nation. Like the Cheyenne River Sioux Tribe, Navajo Nation emphasizes ownership of infrastructure and acquires large government contracts and federal subsidies that support infrastructural build-out and technical training.

The difference between the two is that Navajo Nation emphasizes greater tribal government telecommunications regulation than does the Cheyenne River Sioux Tribe. The Navajo Nation strategy can be characterized by intratribal institutional alignments focused on technical advancement for the nation through ownership and regulation of broadband infrastructure, but not necessarily through ownership of a tribal Internet service provider. Navajo Nation's strategy promotes affordability of Internet services by means of competition and regulation. When I think about the expanse of the Navajo Nation, which is spread across states, I can most appreciate this solution. Like the SCTCA, it must retain flexibility so that it can run fiber and stream signals across such a diverse geopolitical terrain. Moreover, as a tribal nation with a landmass larger than many eastern states, a high number of college and graduate school graduates, and a technical university, with the right leadership in the years to come its members may increasingly be able to position themselves as leaders in digital and design innovation within Indian Country.

CHAPTER 5

Internet for Self-Determination

Practical Needs

If we allow the Indian model to revive and be useful, the disparity between humans and technology will begin to diminish. Then the ability to conceptualize contemporary problems—the environment, the ozone layer, ecology, science itself—can emerge. You would have scientists who come from a more harmonious and balanced sense of who they are as a people.

—Carlos Cordero, "Reviving Native Technologies"

ONE GOAL OF REFRAMING IS TO SHOW THE COMPLEXITIES OF SOCIAL problems that members of a privileged class deem endemic and inherent to reservation life. Reframing means deciding what historical factors shape the background of a problem, and what conditions shape Indigenous possibility within the contemporary moment.

When I began researching the relationship between technology and sovereignty, I was seeking to reframe the way limited access to information and communication technologies in Indian Country had been defined in the information scientific literature. Instead of assuming that Native peoples do not have access to ICTs and asking what factors and conditions limit access, I asked how individuals design and utilize ICTs toward tribal goals. I found that the leaders of ICT projects continuously assess community needs and think about ways of applying technical know-how to meet those needs given conditions shaping the local geopolitical terrain. Yet, visiting Matt Rantanen at the center of operations for TDVnet showed me that I had been missing the forest for the trees. While investigating Native uses of ICTs, I had been looking for a singularly Indigenous approach to uses of ICTs when, in reality, there are as many ways for Native peoples to develop ICT projects as there are Native communities. In a way, that demonstrated colonial thinking on my part. Before, I had been immersed in a body of literature—coming from information science and science, technology, and society studies—that presumed a binary opposite between Native and Indigenous peoples and so-called modern technologically advanced people, a literature written by individuals who had not accounted for the range of knowledge work, both digital and analog, that upholds tribal sovereignty. My instinct was to push back at that false logic. Doing so enabled me to see and articulate a bigger picture.

Tribal uses of ICTs are constrained by the availability of affordable broadband Internet across reservation lands. As more state and federal governments push data online in the drive toward e-governance, tribal knowledge work—the data-driven work used to uphold tribal sovereignty—is constrained by the availability of broadband Internet across reservation lands. Once I realized how this first level of disparity (lack of Internet infrastructure) could shape a second (the inability of tribal leaders to act on information in a timely fashion), I sought accounts of tribal leaders acquiring broadband Internet infrastructures for their reservation communities. I wove these accounts into narrative threads and, from there, visually mapped out the particular strategies each tribe or tribal association engaged in order to acquire a regional broadband infrastructure solution. My goal was to identify the problems that these strategies generate and resolve, as well as the resulting social and political impacts. I wanted to be able to understand the changes that building large-scale information systems introduces into reservation communities. I was returning to this recurring idea: that infrastructures are the crystallization of institutions, and that institutions emerge from the human relationships that form around common work goals.

Imagining weaving against a loom, I was constructing a framework in which the selvage threads consist of a select combination of theories from Native and Indigenous studies and from information science. Native and Indigenous studies teach us that colonialism is an overwhelming social force composed of multiple, overlapping government-sanctioned colonization programs. These programs usually consist of efforts to remap and shape the terrain in order to meet dominant nation-state goals; articulate indigenous peoples as labor or eradicate them as objects of terror; identify a state language by marginalizing all others; and channel acceptable knowledge through authorized institutions while censuring threatening epistemologies. Colonization programs, such as the diversion of waterways in southern California, have a tail—a legacy effect—that continues once the official program (i.e., dam construction) ends.[1] In this way, colonialism is cyclical, and it is very much influenced by government policies, institutional rhythms, and infrastructural capacity.

It is very difficult to describe and explain the impact of colonialism on reservation life. Indeed, one of the effects of colonialism is that, in the United States, members of the non-Native privileged class cannot see or grasp the travails of Native peoples. The concept of a pervasive American Indian genocide is epistemically incongruent with US settler ways of life, thinking, and problem solving. Incommensurable with US pragmatic techno-scientific norms, the idea that Native Americans also require and demand access to high-speed Internet continues to be a challenging concept among computer scientists, network engineers, and human-computer interaction researchers, many of whom are so focused on advancing interface and system design toward entrepreneurial start-up goals that they fail to realize that many people in different parts of the United States still lack the level of access necessary for basic public safety, creativity, and productivity. This epistemic blindness is recognizable in the National Broadband Map, which includes the political boundaries of states and counties but not of sovereign tribal lands.[2] These sovereign lands are largely invisible on the coverage map and therefore invisible to policy makers and network technicians. Only recently—nearly seven years after the Federal Communications Commission released the online map—have policy makers begun to demand visual representation of reservation boundaries on the map. Interinstitutional efforts still need to be made among tribes and industry partners to figure out what sources of data can be used to fill the gaps in the map, and to what degree of granularity.

In the United States, the national telecommunications infrastructure is one of the only major infrastructures—the others are the interstate transportation system and electric power grid—that has been primarily market-driven. The net effect is that urban and semi-urban locations, and locations neighboring major highways, receive the most robust and competitively priced ICT services—landline, wireless, fiber-optic cable, and satellite connectivity—while many tribes lack the infrastructure for basic phone service. Broadband infrastructures are quite costly to construct. With a price tag in the millions of dollars, a steep learning curve for technicians, and the need for long-term business planning in economically marginalized tribal communities, the decision to invest in building out a network backbone represents a significant commitment for a tribe. It also represents a high degree of resiliency and resourcefulness. It shows that tribal leadership really must conceptualize broadband Internet infrastructures as a solution to complex community problems—including long-term economic development planning—well beyond simply gaining access to the Internet.

Thinking about tribal broadband infrastructures as solutions and the resulting impacts of their build-out reminded me of the purpose of reframing. Many times, social problems that are deemed "Indian problems"—including lack of Internet and ICT access in Indian Country—are considered as such due to both a misunderstanding of how colonialism works and misapplied generalization. Every tribe is bounded by a unique historical experience of colonization and a particular relationship within the landscape. Thus every tribe that identifies broadband infrastructure as a workable solution will innovate a unique strategy for acquiring and advancing broadband infrastructure. Only certain aspects are generalizable. In order to identify the problems that tribal broadband infrastructures generate and resolve, as well as the resulting social and political impacts with regard to sovereignty, I cleaved to the original problem that each project leader identified as the reason for investing in broadband infrastructure for his or her community. In this way, I culled those aspects that are particular to each tribe from those that ground a greater narrative about the conditions shaping US tribal access to the Internet.

It is important for Indigenous scholars to pay attention to this, because it forces us to rethink assumptions about the zero-sum agonism between a disenchanted techno-scientific hegemonic world order that inherently clashes with more holistic Indigenous reverential ways of knowing.[3] Acknowledging the amount of work that conscientious and committed Native and Indigenous leaders have done in wiring their communities for the purposes of tribal self-determination, cultural sovereignty, and pathways to decolonization forces us to acknowledge the multiplicity of ways that knowledge moves—the possible channels through which information and data flows—in Native and Indigenous communities, and to what ends. From a tribal perspective, it forces us to think about who we are as a people, the edges of our tribal orientations to truth and change, our potential for acts born out of false consciousness—especially in accord with US techno-scientific infatuation—and our greater goals of decolonization, reducing the authority of settler oppression in our everyday lives. As Anishinaabe scholar Margaret Noori writes, "If Nanabozhoo were among us (and he might be) working to keep the [Anishinaabe] language alive, he would be a hacker, a gamer, half-human, half shape-shifting avatar."[4] Understanding the impacts of Internet build-out in tribal communities reveals that the sovereign acts of tribes are occurring not only through the instantiation of network backbones but also through the digital communication channels—platforms, applications, collaborative work groups—that these network backbones open up and also shows Native peoples' choices about when to use them. It forces us to acknowledge that what are for us key exercises of decolonization—connecting with Indigenous relatives and colleagues, writing to one another, and reclaiming and revitalizing the sources and modes of Native knowledge—are dependent to some degree on the work of network administrators keeping reservation communities online. If we are, as Cordero writes, to revive Native technologies—Native science and methods based on Native ways of knowing—we must acknowledge this digital and technological aspect of our labor, our contemporary historical moment, and, reaching through this, arrive at a harmonious sense of who we are and how we know in the contemporary moment.

TDVNET: ADVOCACY, SUBSIDIES, SPECTRUM

The leaders of the Southern California Tribal Chairmen's Association framed a tribal broadband network as a solution to the need for community Internet connectivity, economic development opportunities, and the means of sharing cultural knowledge and providing youth with access to online educational tools. To provide community Internet connectivity, TDVnet project leaders created community media labs and classes and established Southern California Tribal Technologies as an affordable wireless Internet service provider for tribal residents and neighbors. To create economic development opportunities, TDVnet project leaders conducted Cisco network training classes and hired tribal members and local Native residents to work on SCTT projects and at the spin-off enterprise Hi-Rez Digital Solutions. To share cultural knowledge, TDVnet project leaders created the Tribal Digital Village archive, where tribal members post photos, videos, language lessons, oral histories, and other cultural knowledge. To support educational objectives for youth, TDVnet project leaders provided free access for schools, libraries, and community centers across the nineteen tribes. If we understand broadband infrastructures as consisting of devices, systems, content, policies, and the people who make the whole network function, we can conceptualize the changes TDVnet introduced within the service community across these dimensions.

With regard to devices, building out the TDVnet backbone introduced project leaders to the cycle of testing and modifying devices to fit connectivity needs and environmental constraints. The lack of electricity at some of the mountaintop base stations had construction crews hauling generators and battery packs via four-wheelers and helicopters. To prevent system outages due to loss of power, TDVnet backbone crews innovated a more sustainable electric power solution; they began wiring solar panels and small wind turbines to the generators and battery packs. They also began testing and comparing wireless dishes in an effort to find those that could stream the greatest amount of bandwidth through the hill and valley terrain while sucking up the least amount of electricity. Much like the KPYT-LPFM radio station engineer bending the antennae to provide a stronger signal over a greater expanse of the Pascua Yaqui reservation, TDVnet engineers were adjusting the dishes beyond their original design. They also began testing and innovating new devices, such as wireless mesh transmitters, for their canyon communities and other communities located out of the line of sight of the towers.

With regard to systems, TDVnet programmers worked with community members to articulate an ontology for the Tribal Digital Village archive. As more people began connecting to the Internet from their homes, the demand for bandwidth grew, and network administrators began redirecting available spectrum. With TDVnet being run almost entirely off unlicensed spectrum, network administrators began thinking of how to gain access to more spectrum across the checkerboard of reservation lands and how to reprogram the network system and devices to broadcast more efficiently.

To TDVnet administrators, the demand for greater bandwidth signified an increased use of heavier streaming content. In part, this has to do with the changing nature of Internet content and the availability of streaming broadband devices. Rantanen told me that, from the late 1990s through the first decade of the 2000s, the tribes had developed a process for handling paper-based government grants applications. But when the Federal Register moved to a primarily online mode of dissemination, tribal departments found that they were missing out on notices of grant opportunities. Since then, the tribes have had to adjust their work modes in order to meet the electronic application cycles. TDVnet allowed tribal departments to move from a paper-based grants application cycle to an electronic application cycle. The next step has to do with increasing bandwidth to handle streaming video content and gaming.

Yet perhaps one of the most compelling impacts of TDVnet—both within the service community and at the national level—has to do with policy changes prompted by the build-out. In seeking funding opportunities for subsidizing access costs, TDVnet leaders discovered that tribal libraries and schools are not eligible for e-rate funds, federal funding designed to subsidize Internet access for public schools and libraries. Part of the reason is that funding is administered through the states, positioning state sovereignty against tribal sovereignty, even though tribes have a direct relationship with the federal government. Rantanen has spoken on this issue in tribal telecommunications forums.[5] He and others have also advocated for changes to the FCC's policy on allocating spectrum over sovereign tribal lands.[6] The complicated work of obtaining rights-of-way to construct broadband infrastructure on sovereign tribal lands has also prompted discussions about adjusting this federal grant requirement. It is within the sovereign rights of tribes to administer property inheritance and regulate domestic relations according to customary practices. Tribal home owners may not have deeds or documentation other than oral history or local knowledge of family lineage. This can make it nearly impossible for many tribes to comply with the federal requirement for obtaining rights-of-way in tribal lands, especially when it comes to stringing fiber-optic cable. Even with right-of-way documentation, tribes still need to work with elders and community leaders on obtaining permission to build out infrastructure in a manner that is respectful of the landscape.

Rantanen described a moment in which the SCTT build-out team had to delay plans to lay a tower foundation because of the natural growth of the manzanita plant in the same area. Manzanita is a sacred plant for the Kumeyaay peoples; elders and traditional historic preservation officers reminded network technicians and construction crews that they could not cut down the manzanita to make way for the concrete foundation and fencing. Interestingly, a seasonal wildfire burned across the hillside, sweeping away the manzanita in the proposed build-out location. At that point, community leaders gave the SCTT team permission to lay the foundation and set up the fencing. This example shows that while construction teams must comply with a standard environmental impacts assessment requirement, working with tribes also requires understanding customary laws about setting up infrastructure—constructing towers, digging for fiber-optic cable, stringing aerial fiber, clearing roadways, and hosting blessings—within the homelands.

Between modifying devices, upgrading systems, teaching community courses on digital content creation, and policy advocacy, TDVnet personnel have developed new skill sets as certified tribal broadband professionals informed by an awareness of Internet access as a sovereign tribal right. At present, the Zero Divide Foundation—a digital rights advocacy nonprofit based in San Francisco—is conducting a community impact evaluation of TDVnet services. Yet the national impact of TDVnet leaders is noticeable. Rantanen advises the FCC Office of Native Affairs and Policy, is active in Telecommunications Forum of the National Congress of American Indians and presents on tribal broadband issues at TribalNet and such venues as the Tribal Telecommunications and Technology Summit. The work of deploying TDVnet has also informed the visionary 2006 document of the Indigenous Commission for Communications Technologies in the Americas.[7] All these changes have occurred in little more than a decade, beginning with nineteen tribes with a colonial legacy of being disconnected and having no, limited, or costly Internet service, to TDVnet leaders setting the agenda in national tribal telecommunications forums.

RED SPECTRUM COMMUNICATIONS: CREATING DEMAND ENCOURAGES INVESTMENT

In the first decade of the 2000s, Valerie Fast Horse and Tom Jones framed a tribal broadband network as a solution to the unmet need for connectivity and the means of supporting expressions of cultural sovereignty. Establishing Red Spectrum Communications as an Internet service provider and wiring tribal administration buildings and community computing centers addressed the need for connectivity. Supporting Rezkast and land management efforts through a global information system positioned Red Spectrum to support expressions of cultural sovereignty.

The Coeur d'Alene Tribe has upgraded devices continuously, with IT personnel wiring tribal administration buildings for a local area network, to network administrators increasing bandwidth for hosting Rezkast, to construction crews attaching wireless dishes and positioning towers for a wireless Internet service provider. Now, subcontractors are in the process of stringing fiber-optic cable across 275 miles of forested river valleys and flatlands for the Coeur d'Alene fiber-to-the-home plan, which includes the hardware, digital devices, tools, and vehicles to support the long-term $12.3 million build-out.

Fast Horse and Jones made the decision to upgrade to fiber optics based on community use data that showed reliance on heavier content not only for leisure and entertainment purposes but also to support tribal administration, such as within the tribal global information system. As project leaders, Red Spectrum personnel will upgrade the network systems to meet the demands of fiber-optic use. Tribal chairman Allan describes the goals of the fiber-to-the-home plan as creating opportunities for educational access, economic opportunity, and job growth.[8]

With regard to policy, when asked to describe the fiber-to-the-home plan, Fast Horse harnessed the language of self-determination: "True economic development won't happen if we only focus on our financial capital while ignoring the human spirit. Our challenge is to revitalize the spirit of our people through true self-determination. It is our hope that by lighting up the reservation with a fiber optic network we will spark our most creative minds and encourage the knowledge-based economy we've been striving to develop."[9] Approaching tribal broadband from the perspective of self-determination and cultural sovereignty is part of the reason Fast Horse represents such a strong voice in forums such as the NCAI Telecommunications Forum and through the FCC Office of Native Affairs and Policy. Red Spectrum leadership affects broadband policy at the national level. With the new fiber-optic plan in motion, Red Spectrum operations are expected to increase and expand, creating jobs, training opportunities, and community demand for greater bandwidth, in short, creating the Indigenous knowledge work that Fast Horse envisions.

The impacts are difficult to comprehend without recalling the intergenerational struggles of the Coeur d'Alene people, who have been contending with more than a century of industrial pollution in the tribal waterways and mountains. The Red Spectrum fiber-optic upgrade is not only about meeting community demand for bandwidth or creating local jobs; it is about preparing younger generations of the Coeur d'Alene Tribe to leverage ICTs toward revitalizing the land, waters, and spirit of the people in spite of ongoing colonization.

LAKOTA NETWORK: TRIBAL KNOWLEDGE WORK ENTERPRISE

A commonly referenced reason for investing in these infrastructures pertains to the need to create economic and educational opportunities for younger generations. This is a key dimension of self-determination that is not often explained. When the US Congress ratified the Indian Education and Self-Determination Act in 1978, for the first time in a long time, tribal leaders were able to design and administer their own social service and other programs on the reservations. Before then, many programs were created by federal agencies composed of individuals unfamiliar with particular community needs—much less the long histories of the peoples within their homeland—especially with regard to the need to create programs to sustain future youth.

This is why concepts like economic self-determination are so powerful in Indian Country. The Cheyenne River Sioux Tribe Telephone Authority was one of the first 100 percent Native-owned telecoms operating within the United States. For years, it has been a source of jobs, training, and affordable telecommunications connectivity on the Cheyenne River Sioux Reservation. Gregg Bourland and J. D. Williams first proposed investing in the Lakota Network as a way of addressing the need for more local knowledge-based jobs and affordable Internet connectivity. The Cheyenne River Sioux Tribe Telephone Authority command of the Lakota Network allowed Bourland and Williams to both monitor its build-out and assess community use and readiness. This level of command prefigured the development of spin-off tribal enterprises—the Internet service provider, credit card billing company, and data digitization company—that would prove to be beneficial in different ways.

With regard to devices, shifting the telephone authority to a telecom and Internet service provider required purchasing the towers and receivers and setting up the computing facilities and hardware for training local technicians. The data digitization function of Lakota Technologies, Inc., weaves industrial printers and scanners into the Lakota Network ecology of devices. With regard to systems, telephone authority technicians underwent the same adjustment that Brian Tagaban described occurring at Navajo Nation; as IT personnel learned about telecommunications standards and network administration requirements, they adjusted internal work systems accordingly.

With regard to content, Cheyenne River Sioux Tribe personnel now handle different kinds of data through different kinds of knowledge work within the reservation. These include supporting streaming content through the Internet service provider, designing tribal administrative websites, and digitizing large amounts of data for the National Library of Medicine and other contractors. Add to that the Cheyenne River Sioux Tribe radio station, which is run out of the Lakota Technologies, Inc., building, and we get a sense of the diverse information flows streaming through the Lakota Network.

With regard to policy impacts, the leadership of the Cheyenne River Sioux Tribe has been an active voice for a while on matters of tribal telecommunications, expressing topics of interest and raising concerns with the Native American Broadband Association and in different tribal telecommunications forums. Being able to represent telecommunications issues from the perspective of tribal economic self-determination allows the Cheyenne River Sioux Tribe leaders to point out critical issues related to broadband in Indian Country. In 2004, the FCC issued an inquiry regarding the possibility of providing broadband over existing power line systems. At the time, the FCC Office of Native Affairs and Policy was not yet formed. Concerned that the federal government might undercut tribal efforts to harness their own broadband business solutions, CRSTTA leaders asserted that "only by consulting with individual tribal governments on a case-by-case basis can the Commission as a practical matter determine whether an Indian reservation is underserved, and also determine the manner in which that tribal government may wish to address the question of availability of broadband Internet services within Indian tribal territory."[10] This proved to be an important point, as the FCC would later acknowledge when FCC chairman Michael Copps would advocate community-based broadband solutions that would best meet tribal jurisdictional limitations and particular community needs. The CRSTTA commentary reminded FCC authorities of the tribal right to self-determination.

Underlying the CRSTTA investment in the Lakota Network is the goal of encouraging tribal members to develop the skills to work in a knowledge economy. Development of a diverse knowledge-based business portfolio has allowed tribal business and political leaders to teach younger generations about the kinds of knowledge work that could be available to them through the reservation. For the Cheyenne River Sioux Tribe, command of the local broadband infrastructure has been empowering in this regard. It is important to note the speed and pace at which these policies are having an effect at the tribal level, considering that the Indian Education and Self-Determination Act was signed just a little over thirty years before.

NAVAJO NATION: LONG-TERM BROADBAND INVESTMENT

In discussing the impact of broadband in Indian Country, perhaps one of the most difficult aspects to explain concerns misunderstandings about the originating need for affordable and robust broadband connectivity. Others have written about the challenges of supporting business in Indian Country. Many of the factors that now preclude many tribes from either launching their own Internet service providers or contracting with competitively priced providers have to do with these economic constraints. Entrepreneurs must engage with federal Indian law, comply with the taxation and regulation requirements of tribes, obtain rights-of-way across tribal jurisdictions, and, in some cases, be willing to abide by tribal hiring preferences. Many entrepreneurs and small businesses do not know how to work with tribes. There are assumptions about reservations as impoverished places that cannot support market demand.

Driving through the Navajo reservation—past miles of red and yellow desert, through scrubland, and low pine forests bordering a handful of small towns—it is easy to think this way. The striking juxtaposition of a Macintosh computer, designed for use in the suburbs of wealthy Palo Alto, California, against the Ganado, Arizona, landscape highlights this divergent thinking. When Native American Access to Technology Program managers identified satellite Internet access as a reasonable solution to providing connectivity across the expanse of the Navajo Nation, they were not fully considering the particular community information needs of the Navajo Nation.

Various ICT leaders within the Navajo Nation worked together to draft the Internet to Hogan plan so that it would specifically address the unique needs and infrastructural possibilities involved in crossing the diverse landscape. The plan's goal was community connectivity over time, with an emphasis on tribal command of the infrastructures and advancement of various tribal administrative ICT goals.

With regard to devices, over time, this resulted in the acquisition of facilities and hardware to support computing labs at Navajo Technical College, wireless towers and devices, and now, a network of terrestrial and aerial fiber-optic cable throughout the Four Corners area. With regard to systems, the Navajo Nation has, within a decade, shifted from administering a series of disconnected local area networks to hosting a robust—wireless, landline, and fiber-optic—broadband infrastructural grid through the Tribal Utilities Authority.

With regard to content, the Tribal Utilities Authority, Navajo Technical College, Diné College, and other workplaces and educational sites across the reservation now host advanced technical skills training, courses, and degrees for tribal members. Diné College offers a bachelor's degree in computer information systems, and Navajo Technical College has become Navajo Technical University, offering bachelor's degrees in computer science, digital manufacturing, and new media and associate's degrees in computer-aided drafting, geographic information systems, and information technology. KTNN, the Nation's AM radio station—and also the last AM radio station in the United States to receive a 50,000-watt supertransmitter permit from the FCC—is now streaming online. In time, the Navajo Nation's KWRK-FM station will also be available online.

Meanwhile, with regard to policy making, Brian Tagaban became actively involved in planning the Tribal Telecom and Technology Summit and advising the FCC Office of Native Affairs and Policy. Over time, it will be interesting to see how the Navajo Nation approach to telecommunications and broadband regulation—innovation and affordability through regulation—shapes FCC policies on providing grants and loans for broadband Internet service providers throughout Indian Country.

Observing the Navajo Nation realize the Internet to Hogan plan teaches us not only about the impacts of intra-tribal planning for broadband connectivity but also about how tribal command of the infrastructure shapes control over regulatory matters and attentiveness to community needs.

COMMON IMPACTS OF TRIBAL BROADBAND DEPLOYMENT EFFORTS

When it comes to deploying broadband infrastructure across tribal lands, project leaders work through a few common conditions. They must account for the inherent, legal/political, and cultural sovereignty of tribes; the technical specifications of devices; sources of funding; policy and regulation at the local, state, and national levels; support and investment of the tribal council and other key partners; and the skill and interest of the target communities. Community needs assessments and pilot ICT projects are key steps in the overall strategy for designing and deploying tribal broadband Internet infrastructures. Project leaders possess a unique skill set, including business, political, and technical acumen. They know the histories of their peoples and engage tribal council and other leaders. They work cooperatively in local and national policy arenas, advocating for the changes they need in order to realize increasingly robust infrastructural deployment plans. They focus on training younger generations, incorporating technical training, digital literacy, and policy awareness into the skill sets of project employees and tribal community members.

As a result, the inherent cultural sovereignty of tribes drives broadband goals, priorities, and next steps, with the legal and political enforcement of sovereignty reinforcing the need for tribal access to licensed and unlicensed spectrum, feasible infrastructural subsidies, and rights-of-way across state and federal jurisdictions in Indian Country. Tribal leaders exercise sovereignty when they assert rights-of-way for constructing towers and laying cables on or near tribal lands. They assert sovereignty when they demand that the FCC create a tribal priority for licensing spectrum, the airwaves that carry digital signals. They assert sovereignty when they tax providers of Internet services to peoples on tribal lands. They assert sovereignty when they regulate infrastructural uses. They assert sovereignty when they remind federal authorities to respect tribal broadband efforts as a matter of self-determination.

Understanding how tribes effectively build and maintain broadband operations reveals the critical importance of tribal ICT champions leveraging partnerships in order to guide tribal, regional, and US broadband Internet policy, access, infrastructure, and utilization decisions. As of this writing, there are at least twenty tribally owned or Native-owned telecommunications and Internet service providers in the United States, with more planning to move in this direction. While this represents a fraction of federally recognized tribes, the political and social clout of the ICT champions affiliated with these entities is remarkable.

The champions highlighted here—Valerie Fast Horse, Matt Rantanen, Brian Tagaban, Greg Bourland, and J. D. Williams—have all served or continue to serve, along with many others, in an advisory capacity to the FCC Office of Native Affairs and Policy and in other spheres, including Native media advocacy groups, community technology advocacy groups, national computing advocacy groups, and Native business associations. As advisers to the FCC, these individuals directly address the areas of Internet governance outlined in the National Broadband Plan and are called to deliver congressional testimony as notices of inquiry are released and consultation efforts prepared. They have had to work through a variety of social, economic, political, legal, and technical barriers to setting up infrastructure for Internet service providers on reservations. They have endured failed attempts and slow starts. It is important to frame failures and slow starts in a positive way, as opportunities to learn how and where to improve operations. Obstacles, failures, and slow starts are not indications of an "Indian problem" or a culture that does not accept technological change but are more than likely outcomes of histories of colonization, legacies of economic policies designed to keep Native peoples from acquiring the capital—social, financial, and political—to build in a self-sustaining manner.

Once we reframe obstacles, failures, and slow starts as design constraints rather than impossible barriers, we can begin to build more flexible systems, designed for the unique landscapes and communities they are meant to serve. Understanding how TDVnet, Red Spectrum Communications, the Lakota Network, and the Navajo Nation network got their start reveals the constraints faced by tribal ICT champions as they attempt to provide wireless Internet for their communities. These cases also teach us about how technological change emerges over time across tribal landscapes, with great intention, at many levels of operation. Processes of visioning, piloting, assessing, and evaluating precede processes of proposal writing, grant making, investment, design and deployment, and use.

Occurring over and over again, in an iterative fashion, these processes contribute to the slow and deliberate weaving of broadband infrastructures and services—tangible and intangible—into Native landscapes. Over time, every tribe will make its own decisions about how to address advances in digital technologies. For tribes, gaining access to broadband is not just about plugging into a grid. It is about establishing partnerships for constructing that grid from the ground up and advocating for the policy changes needed to weave this critical infrastructure into the diverse geopolitical ecologies of Indian Country.

CHAPTER 6

Network Sovereignty

Broadband and the Rights of Tribes

> One of the challenges is that of communication with all those who are fighting for this. Technology should also seek the path of below, so that the weaving of this network can be made visible in the Other Campaign. This is a job for right now.
>
> —Subcomandante Insurgente Marcos, "The First Other Winds"

BEFORE I BEGAN THIS PROJECT, I WAS OPERATING UNDER THE FOLLOWing misguided assumptions. I presumed that most people on tribal reservations were impoverished and, as a result, had no reliable Internet or telecommunications services. I presumed that there was a fundamental zero-sum conflict between tribal traditions and digital design efforts; I underestimated the potential for Native peoples' digital creativity. I assumed tribes were not working together to address information and communication technologies and telecommunications problems and policies in Indian Country. I underestimated the effect of colonial legacies.

This research shows that the process of building out broadband infrastructures is constrained by the sovereign rights of tribes, in terms of both jurisdictional limitations and the underlying reasons for implementing broadband as a solution for community needs. This research shows tribal Internet advocates working together frequently and productively, across domains of industry, government, education, and tradition. This research showed me how policy makers in Washington, D.C., by not considering the rules of tribal sovereignty, develop policies that preclude tribal peoples from creating the programs and services they need for self-governance goals. What I did not expect to find was how the build-out process creates opportunities for project leaders to convey the obligations of federal and state governments, and other project partners, to Native peoples and governments within the context of sovereignty and self-determination. This research has led me to conceptualize the following premises underlying Native uses of ICTs, and, specifically, around the social and political implications of tribal broadband infrastructures.

One is that the histories of ICTs are inevitably intertwined with histories of colonization, sovereignty, and self-determination. The second is that federally recognized tribes, having understood this, now exercise their sovereign rights to acquire broadband Internet infrastructure and services for the peoples living on or near their reservations. Tribal leaders command the build-out of broadband infrastructure as part of a greater effort to enable the self-governance of Native peoples. The third is that the process of network design, deployment, and regulation embeds these infrastructures within the ongoing negotiation of the sovereign rights of tribes within the United States. This has major implications for how design, deployment, and regulation are enacted.

THE THIRD NETWORK: AN AMERICAN INDIAN SOCIOTECHNICAL LANDSCAPE

Understanding the specific ways that ICTs are interwoven throughout Indian Country helps us conceptualize how the histories of ICTs are intertwined with histories of colonization, sovereignty, and self-determination. By way of explanation, I share a story about a visit to a leadership and policy-making forum in Indian Country, the National Congress of American Indians Tribal Leader/Scholar Forum.

In the summer of 2010, I caught a plane with my colleagues in the Indigenous Information Research Group to River Bend, South Dakota. It had been a rough flight coming in over the Black Hills. Thunderheads burned red and orange against the setting sun and grew dark with lightning. The plane bounced through the storm. I exchanged glances with Allison Krebs and Miranda Belarde-Lewis. Allison clasped her hands together. Semi-joking, her eyes widened and she whispered, "They know we're coming," referring to the descent of tribal leaders of ancestral lineages coming from all parts of Indian Country to this place for sharing knowledge. These dimensions accompany convenings of Native leaders: the acknowledgment of the environment, the Indigenous histories of places, and the reverence for ancestors.

In the shuttle from the airport to the hotel, we discovered that Miranda was friends with a coworker and relative of another passenger in the van who was also on the way to the forum. Allison knew another person in the van. We passed a billboard that read "Don't drink and drive. It's uncivilized." Raindrops streamed like long hair against the vehicle windows. Lightning lit the dark plain.

The next morning, the conference opened with the words of elders. A woman spoke on the need to reclaim the sovereignty of the airwaves over tribal lands. Tribes need to share their knowledge, news, and language through the radio and other means. She described a future in which tribes could launch their own communications satellites through their own airspace. She asserted that airwaves, like waterways and forested lands, should be held by the United States in trust for sovereign, recognized tribes.

During our presentation on the subject of tribal ownership of information and data, a leader from one of the northwest tribes reminded us, gently, not to forget the treaties. I thought about that admonition and about the role of information as surveillance in the long history of colonization among military authorities, settlers, and Indians.[1] I thought about Eloise Cobell's painstaking accounting, which identified all the mathematical mistakes, technical errors, and misreading of regulations by US federal authorities and proving the theft of billions of dollars and hundreds of acres of lands from Native and African American farmers.[2] I thought about how the treaties were so often written in English, not the Native languages, on paper—the skin of trees felled on Native lands—and locked in US government archives, far from the lands and leaders who negotiated their original intent. I thought about how, in spite of the language of the treaties, generations of Supreme Court justices had favored the immoral and unjust doctrine of discovery over their understanding of international treaty making as an acknowledgment of the inherent sovereignty of peoples.[3]

At that particular convening in the township at the foot of the Black Hills, I learned that there is the policy of sovereignty, there is the material substance of it, and there is its enactment. When treaties were documented on paper, the paper was the materiality of it. Treaties, policies, and court precedents incorporate Native lands, waters, and bodies into the legal system underpinning US law and policy. The rights are not disconnected from the substance. As tribal leaders negotiate and assert the rights of Native peoples across digital media, the supporting digital devices and infrastructures become imbued with the values of sovereignty. The policy, materiality, and enactment of tribal sovereignty exist within the dynamic flux of human relationships.

For years, tribal broadband and telecommunications advocates J. D. Williams, Matt Rantanen, Valerie Fast Horse, Shana Barehand Green, Traci Morris, and many others have been voicing their concerns to members of Congress and the Federal Communications Commission about the lack of critical telecommunications and broadband infrastructure in Indian Country. In 2010, FCC chairman Julius Genachowski appointed Geoffrey Blackwell, Muscogee (Creek), to head the newly formed Office of Native Affairs and Policy.[4] In 2011, the authorities of that office began traveling to different places in Indian Country, hosting workshops on federal broadband and telecommunications initiatives designed to assist rural and tribal borrowers, and learning about the issues involved in building out a durable ICT infrastructure through Indian Country.

In 2012, they walked through a reported broadband wireless zone on the lands of the Confederated Tribes of the Colville Reservation bordering northeastern Washington. Sections of disconnected and unused fiber-optic cable lay by the side of the road. They saw that areas reported as covered in the National Broadband Map lacked basic infrastructure.[5] They observed firsthand that federal policies were not aligning with the geopolitical landscape of Indian Country. They observed that digital devices did not function in areas lacking infrastructure, that the infrastructure was not placed to function on top of the mesas, as in Hopi, alongside canyons for the Hualapai, through tree lines for the Yurok, amid deep freezes and heavy snows, as in the North Slope of Alaska, or across the checkerboard of the southern California reservations. They heard from tribal traditional historic preservation officers, who expressed their concerns about telecommunications providers building towers on sacred sites, without permission from tribal elders or other authorities. I recalled how, during the summer of 2011, as Mike Wilson and I were leaving the Tohono O'odham reservation in southern Arizona, I spotted telecommunications towers atop Kitt Peak, a sacred mountain, a center of Creation, for the Tohono O'odham people, and the location of Kitt Peak National Observatory, a National Optical Astronomy Observatory, funded by the National Science Foundation. I wondered if credible representatives from the National Science Foundation, the National Optical Astronomy Observatory, or the regional telecommunications providers had consulted with the Tohono O'odham elders, traditional historic preservation officers, or elected tribal leadership before constructing these large-scale systems on sacred sovereign land.

Technological devices are woven throughout Indian Country. There are vehicles, two-way radios, AM/FM radios, televisions, push-button telephones, smartphones, and laptops. In some places, there is the infrastructure to make these work: gas station pumps, server rooms, phone lines, telecom towers, solar panels, power lines, and wind turbines. Leaders in Indian Country use these devices to defend and enforce the sovereign rights of tribes. Policy making is about communicating. Indian Country is, as a matter of speaking, the legal definition of a place wrought by the overlapping colonial policies of conquest and treaty making, removal and relocation, allotment and assimilation, reorganization, termination, and, now, self-determination.[6] There are the devices, there are the people, and there is the policy as it is negotiated over time. The relationships between these three emerge against the backdrop of the geopolitical landscape.

Broadband leaders in Indian Country understand the sociotechnical implications of the tribal command of broadband infrastructures and services. In 2009, members of the Native American Broadband Association described broadband as the third network, a network critical to tribal integrity.

> Twice before networks made major changes to Indian Country. In the 19th century a network of railroad tracks were laid across Indian Country resulting in the wiping out of the buffalo herds, providing quick movement of military personnel, and the influx of millions of immigrants. Most of the railroad roadbeds laid down then are still in the same place today over 100 years later. The second network was the system of electrical and telephone lines laid out in rural America in the 20th century. Once gain, once the lines were put in they tended to stay where they were. . . . For tribal government however, broadband service will play a vital role in nearly everything they do. . . . These broadband networks to be most effective need to be shared with neighboring towns and areas off the reservation. . . . While making sure that networks inter-operate, tribes need to protect their tribal sovereignty in the data information age.[7]

This expression—broadband as the third network—highlights an important dimension of these infrastructures: they bear a colonial legacy and a future impact that is closely tied to the history of Native peoples in the United States.

This meaning is captured in the content flowing across the networks. Tribal personnel transmit information and data transmitted across health databases, accounting systems, fax machines, and via e-mail. There are the ways of knowing that elders and leaders share with younger generations, in person while in the homelands and now, via Rezkast, by streaming radio and podcasts. Content pertinent to the integrity of the tribe relates to the homeland, and by the word *homeland*, I am not referring to a geographic point on a map such as the National Broadband Map. By *homeland*, I am referring to the histories of beings living in right relation to one another within a landscape slowly shifting over the course of hundreds of human generations.[8] The Native peoples of the United States will not leave or sell their land just because it lacks the infrastructure for Internet connectivity. As one Supreme Court justice stated in 1960: "It may be hard for us to understand why these Indians cling so tenaciously to their lands and traditional tribal way of life. The record does not leave the impression that the lands are the most fertile, the landscape the most beautiful or the their homes the most splendid specimens of architecture. But this is their home—their ancestral home. There they, their children and their forebears were born. They, too, have their memories and their loves. Some things are worth more than money and the costs of a new enterprise."[9] At their core, tribal governments are composed of people working through institutions they have struggled to articulate in order to protect and revitalize the homelands. It is an ongoing struggle, and not without critique and the need for meaningful change.

As an Indigenous information scientist, I see Indian Country as an assemblage of devices, policies, and institutions emerging out of Native peoples' unique historical and ongoing relationships with one another, with neighboring non-Natives, and within the homelands. These relationships are characterized by the qualities of information and knowledge that beings perceive, share, communicate, and experience among themselves. Ways of knowing shape rules and behaviors. Individuals work through institutions to codify and create forms of information and knowledge. Institutions require infrastructure to function, and at this time in Indian Country, robust and durable broadband infrastructures are integral to the functioning of tribal governments.

INDIGENEITY AND SELF-GOVERNANCE IN A NETWORK SOCIETY

The second idea that has emerged from weaving together the narratives of tribal broadband deployments is that federally recognized tribes exercise their sovereign rights to acquire broadband Internet infrastructure and services for the peoples living on or near their reservations. Tribal leaders command the build-out of broadband infrastructure as part of a greater effort to enable the self-governance of Native peoples. At this point in history, the sovereign rights of tribes and the policy of self-determination are mechanisms toward this greater effort.

In his 1997 characterization of the network society, Manuel Castells forecast that as ICTs became locally available, identity-based groups would utilize these tools to organize politically and communicate their socially exigent status to a global audience.[10] He theorized the rise of identity-based communalities amid a globally interconnected network society. Yet it is important to remember that the forms of political engagement utilized by Indigenous peoples—and especially with regard to sovereignty in the United States—are not identity-based movements but in fact represent long-term flexible modes of governance for land-based peoples. I describe this distinction by providing a framework for understanding Indigenous peoples' uses of ICTs within the context of an imagined global network society.

First, we must acknowledge the political roots of Indigeneity and that, as Ron Niezen asserts, Indigeneity cannot be understood without reference to the governmental power and technological advance of modern nation-states.[11] During the era of discovery, conquest, and treaty making, it was the practice and habitus of European settlers to classify the indigenous peoples of non-European lands according to the perceived complexity of their tools. Enlightenment-era European ethnographers ranked the indigenous peoples of the Americas as lesser than the indigenous peoples of Africa, who were perceived as being effective only as labor, and greater than the indigenous islanders of the Pacific, who were perceived as idle.[12] These perceptions, written down and codified into social policy, would come to inform the maltreatment and misunderstanding of Native peoples of the Americas and all others on a non-European, and therefore non-enlightened, scale.

Second, we have to acknowledge that the US history of the colonization of Native lands, bodies, and waters has shaped the ability of Native peoples to socially organize and communicate. Armed with a sacrosanct belief in the correlation between moral progress, private property, and technological advance, settlers purveyed public and Christian schooling, surveyors' tools and mapmaking, dam construction, telegraph communications, agricultural work, factory work, and the internal combustion engine as means of settling indigenous lands, waters, and bodies in the Americas.[13] While Supreme Court Justice John Marshall was utilizing *Worcester v. Georgia* to define the domestic dependency of the Native peoples of the United States—precolonial wards of a technologically advancing state—Samuel Morse was refining the electromagnetic telegraph and Morse code.[14] Thirty years later, the Pacific Railroad Acts abetted the removal and relocation of, and warfare against, Native peoples whose lands were tied up in the build-out of the railroad and associated logging and

FIGURE 6.1. The gestalt of Manifest Destiny permeated the settler imaginary during the era of US westward expansion. In this painting, the artist depicts the "spirit of the frontier" as a European woman with yellow hair, holding in her right hand a school book and in her left telegraph wires, as she leads settlers toward the frontier. The artist used darker shadows and hues to depict Native peoples and wild animals fleeing from the settlers' trains and caravans. The image was widely distributed in travel guides at the time. John Gast (1842–1896), *American Progress*, 1872; chromolithograph, 12¾ × 16¾ in. Prints and Photographs Division, Library of Congress.

mining enterprises.[15] Telegraph operators worked with railroad entrepreneurs to set up telegraph posts at each railroad station, so that military authorities could be informed of perceived movements in Indian Country.

The nineteenth-century painting *American Progress* captures this gestalt (fig. 6.1). George Croffut, a publisher of travel guides, commissioned artist John Gast to depict American Progress, which Croffut later printed in a series of travelogues. Croffut described the idea of progress as a woman "floating westward through the air, bearing on her forehead the 'Star of Empire.' . . . She carries a book—common school—the emblem of education and the testimonial of our national enlightenment, while, with the left hand she . . . stretches the slender wires of the telegraph that are to flash intelligence throughout the land."[16] Depicted in the lower left corner, Indians flee westward under a dark cloud, trailed by the telegraph and railroad. At the time of the painting, Native peoples of the southeast were enduring removal along the Trail of Tears, the people of the Plains were being confined by military force to reservations, federal authorities and missionaries were kidnapping children and sending them to boarding schools, and in the burgeoning US-Mexico borderlands, former Confederate soldiers were being paid government money for Indian scalps and captives. The Colt revolver was a key technology of this era.

By the era of allotment and assimilation, church rolls, government accounting systems, and libraries of court proceedings represented systems integral to the subjugation of Native peoples of the United States. Their original names misrepresented in these systems and the original treaties ignored, the indigenous inhabitants of Turtle Island became codified into an exceptional subclass of American citizens: Indians, or American Indians. Native peoples were prohibited from speaking their languages. There are many accounts of teachers abusing Native children for speaking their languages while in school. We can relate the era of allotment and assimilation—with its industrial and agricultural job training programs, forced schooling, theft of Native lands, and prohibition on languages and spiritual practices—to a particular phase of colonialism in which indigenous bodies are inscribed as labor or eradicated as objects of terror, indigenous languages are marginalized in order to sanction a state language, and government-authorized institutions build systems that channel useful knowledge while censuring indigenous epistemologies.

Finally, in order to understand Native peoples' uses of ICTs within the context of an imagined global network society, we must acknowledge that Native peoples have organized to protect their homelands in spite of waves of modern nation-state colonization. Native youth learned from their parents' experiences of removal, relocation, allotment, and assimilation. Communicating in the privacy of living rooms, while hunting, on the edges of ceremonial grounds, during work hours, or while serving in the military, Native peoples share their experiences of colonization.

Linda Tuhiwai Smith writes about this dimension of Indigenous peoples' knowledge. It is not confined to an ecological awareness of the homelands; it also incorporates knowledge of how to survive the worst abuses of colonization.[17] It includes lessons of resiliency and endurance. Regardless of where they are from, Native and Indigenous peoples share the experience of nation-state methods of colonization—most often couched in the terms of the economic imperative to industrialize—that binds a highly politicized awareness of what it means to be Indigenous. Now a global and supranational political expression, the mobile and diffuse self-governance of Indigenous peoples got its start from the localized political engagement of Native and aboriginal peoples of different countries undergoing phases of severe colonization. At their core, the mechanisms of colonialism are designed to curb the ability of Indigenous peoples to communicate with one another, mobilize, and re-create political institutions for reclaiming occupied terrain.

During the post-1960s era of self-determination, the telecommunications industry in the United States was characterized by aggressive free market capitalism, resulting in monopoly ownership of infrastructure and service providers. By the time the Telecommunications Act of 1996 allowed for the entrance of newcomers—and especially Internet service providers—through a regulatory process, Native peoples were only twenty years into setting up tribal programs independent of federal government authorization and control.[18] In the following year, House Democrat Bill Richardson of New Mexico introduced HR-486, the Native American Telecommunications Act of 1997, amending the Communications Act of 1934 such that the FCC would be required to engage with American Indian and Alaska Native tribal governments and identify ways of promoting the development of infrastructure and spectrum licensing on and around tribal lands.[19] Unfortunately, the bill failed on the grounds that it would have required the FCC to "promote the exercise of sovereign authority of tribal governments over the establishment of communications policies and regulations within their jurisdictions," an effort that, at the time, they were unprepared to uphold, in particular for fear that tribal rights to spectrum licensing and regulation would impede the objectives of the Telecommunications Act of 1996.[20] It is no coincidence that the discourse of Internet entrepreneurship is marked by the discourse of Manifest Destiny.[21] Consider the terms and phrases *information wants to be free*, *Electronic Frontier Foundation*, and *Internet pioneer*. For Native peoples, it is as if the imperial urge to westward expansion moved into the cybersphere. Understanding the contemporary ICT landscape in Indian Country with regard to the governmental power and technological advance of the United States reveals the roots of factors that continue to preclude Native peoples' acquisition of sovereign control over their own telecommunications and Internet service provision efforts.

Consider the tension between assertions that information wants to be free and that tribes need to claim sovereignty of the airwaves. Imagine what it must have been like for members of the Native American Broadband Association to review federal broadband subsidy programs designed with tribes in mind. "A century ago they wanted us dead," they might have thought. "Fifty years ago they wanted us to be quiet and assimilate. Now they are asking us to help build a major national communication infrastructure across some of the most remote lands within US borders. What new era is this?"

From a nationalist perspective, Indigenous peoples' political expressions—most often made visible to a global audience via media channels—may appear to be disparate, fragmented, ephemeral, and based on the politics of identity. But this is not the case. To be Indigenous is to be a member of a community with a centuries-long relationship within a living landscape. Out of this relationship emerges a particular spirituality and a particular philosophy of self-governance. While Indigenous peoples from around the world relate to one another politically on the basis of a shared experience of modern colonization, the many Native peoples of the United States relate to one another on this point and with regard to the responsibility of caring for their lands and the peoples who live there. What appear to be disparate and disconnected uses of ICTs actually represent a gap in the published literature about Native peoples' means of communicating across divergences effected by generations of colonization programs. Native peoples have many ways of knowing, and in the past thirty years or so, ICTs have been a means for sharing these ways of knowing and negotiating intertribal and tribal-federal policies toward the treatment of Native lands and waters. Commanding the build-out and uses of their own broadband infrastructures allows tribes to also build t information systems, policies, and programs that meet tribal self-governance goals.

Understanding the specific struggles of Native telecommunications and Internet advocates here in the United States helps us also understand that the pathways to tribal network build-out in Canada and Mexico, and in many other countries, likewise bear their own material and political challenges. It is for this very reason that article 16 of the United Nations Declaration on the Rights of Indigenous Peoples demands Indigenous rights to produce Indigenous media and pursue private media ownership as opposed to state-owned media channels.[22] National policies and practices related to Internet access, ownership, censorship, rights to freedom of expression, industry partnership, innovation, and regulation vary in each country, as do national policies and practices specific to Indigenous peoples' rights.

From the user perspective, it is most desirable to imagine that Native and Indigenous peoples across North America can easily connect with one another through mobile devices, but it is important for the most well-connected Indigenous activists, theorists, and entrepreneurs to remember that basic connectivity continues to be an issue in many parts of the Indigenous world. Furthermore, different Native and Indigenous communities in different parts of the United States calibrate their uses of the Internet in different ways, so while one community may be very open to teaching tribal languages through digital tools, another may forbid language digitization. While a US tribal environmental activist group may live-tweet protests, post photos, and disseminate speeches and manifestos, an Indigenous environmental and political rights groups in Mexico may be suffering under an oppressive government regime practicing censorship, activist kidnappings, murders, coercion, and internal information leaks. It is for this latter reason that Subcomandante Marcos and the Zapatistas framed Indigenous uses of communication technologies as a radical act, with the network of decolonial "Otherly" thinkers strengthened through available stable communication channels. In the middle of the first decade of the 2000s, Subcomandante Marcos reminded well-connected allies in urban centers and elite institutions that "technology should also seek the path of below, so that weaving this network becomes visible in the Other Campaign."[23] In some tribal communities, a tribal radio station is more critical, affordable, and immediately useful than a high-speed Internet connection, and in others, a data center enterprise and Cisco network certification for all tribal engineers constitute the next logical step. Paying attention to the social shaping of technologies in Indigenous contexts means acknowledging the particular social objectives of different groups at different times, grounding their particular values and discourses around uses of digital technologies, and understanding the technical limitations and affordances of systems within particular geopolitical terrains.[24] Over time, information scientists, computer scientists, Internet sociologists, and so forth, certainly will generate more nuanced concepts than the Castellian network society, but as Indigenous, Otherly, and decolonial thinkers, it is our job to write Indigenous experience into those sociological formulations, such that the techno-scientific thinkers in our universities and labs do not forget or diminish the colonial dimensions or decolonial potential of ICTs.

NETWORK SOVEREIGNTY AND THE RIGHTS OF TRIBES

The process of broadband network design, deployment, and regulation embeds these infrastructures within the ongoing negotiation of the sovereign rights of tribes. To understand this, we have to examine the ways in which the build-out of tribal broadband infrastructures intersect with the sovereign rights of tribes. Federally recognized tribes exercise the following eight legal-political rights: the rights to self-govern, determine citizenship, and administer justice; the rights to regulate domestic relations, property inheritance, taxation, and the conduct of federal employees; and the right to sovereign immunity.[25] Native peoples also exercise cultural sovereignty: they share information among themselves and with other Native peoples in order to strengthen knowledge of the homeland, histories, Native languages, spiritualities and ceremonial cycles, and ancestral lineage. While these legal/political rights have been negotiated in US courts and through treaty-making processes over several generations, the right to cultural sovereignty emerges from the will of Native peoples.

The first, and perhaps cornerstone, right pertains to the right of tribal peoples to self-govern. While self-governance refers to the cultural sovereignty of a people, or precolonial modes of governance, it also refers to modes of self-governance in the context of domestic dependency within the United States. We must recall that the present colonial arrangement is based on the policy of self-determination: federally recognized Native peoples should and can determine the course of their own social services and civic arrangements within the federally recognized boundaries of their land.[26]

The principle of self-determination plays out in the ways tribes choose to deploy their own broadband infrastructures or negotiate more affordable solutions based on existing commercial infrastructures. It plays out in the way project leaders frame the need for broadband infrastructures, whether for purposes of strengthening cultural sovereignty, economic self-determination, administrative effectiveness, or educational opportunity.

The principle of self-governance plays out in project planning. For example, tribes who do not have the capital to invest in setting up their own Internet service provider must first accrue sufficient credit. For tribes without major gaming operations or other sources of income, borrowing against trust and fee land is one way of acquiring credit for loans. However, this kind of investment takes long-term strategic planning among tribal leaders and with community members, who must come to an agreement about appropriate uses of the land. It is a matter of self-governance to decide in what ways to leverage the tribe's natural resources.

The practice of self-governance also plays out in the ways tribes decide to approach costly network upgrades and technical training. Part of the Navajo Nation's Internet to Hogan strategy includes continually investing in local technical skills training through Navajo Technical College. Likewise, part of the Cheyenne River Sioux Tribe Telecommunications Authority strategy is to advance the local skill set by setting aside a portion of profits to invest in community computing centers and job training for Lakota Technologies, Inc., employees. In both these cases, community and business assessments are key to planning for upgrades, build-outs, and technical skills certifications. These approaches are markedly different from the Gates Foundation's Native American Access to Technology Program, in which non-tribal teams granted the hardware and set up the local area networks but did not partner with local community members in providing a long-term technical training or upgrade and maintenance plan. Ultimately, the program was a learning experience for both tribal participants and program officers as to the magnitude of the planning required. Recognizing that maintenance of tribal broadband infrastructures represents a long-term investment for tribes also means recognizing the sovereign right of self-governance in that tribal leaders are best positioned to make decisions about the feasibility and sustainability of these large-scale infrastructures.

Another key right of tribes is the right to regulate domestic relations. This means asserting jurisdictional boundaries and setting up systems and policies for managing domestic relations within tribal lands. The right to regulate domestic relations pertains to the strategies tribal leaders employ to ensure that state and regional infrastructural plans benefit or do not impede tribal infrastructural plans. As of this writing, the Navajo Nation Tribal Utility Authority, Gila River Telecommunications (Gila River Indian Community), and leadership from the Hopi Tribe are working with the Digital Arizona Council to make sure that state broadband infrastructural plans align with their tribal broadband infrastructural plans.[27] This highly coordinated effort requires joint broadband mapping projects, shared visioning, and attention to future build-out goals.

Where neighboring tribes, states, municipalities, and local Internet service providers do not work together, the potential for acrimonious relationships can contribute to spectrum squatting, refusal to permit construction, development of noncompetition clauses that may enable predatory business practices on tribal lands, and, in general, poor buy-in for long-term investment and cooperative planning. In the late 1990s, a broadband consulting firm ranked Montana as one of twelve states that would be most costly to wire.[28] The result is understandable, considering that Montana has large rural expanses, seven reservations, no big cities, and a largely low-income population. The vision work toward broadband would take some doing. In 2009, Bresnan Communications, Inc., applied for federal funding to construct a middle-mile infrastructure connecting rural communities and sovereign tribal lands in Montana.[29]

Local, smaller Internet service providers objected, pointing out that the Bresnan Communications middle-mile plan would replicate services they were already providing to select rural and reservation communities and could not carry, as was claimed, the most costly last-mile broadband Internet service into the more remote areas. Meanwhile, members of some Montana tribes countered, asserting that their reservation communities were not being served very well by the smaller service providers.[30] Ultimately, federal grants and loan program officers denied Bresnan Communications' $70 million bid. In part, these kinds of failed efforts result from lack of coordination among stakeholders early on and, with regard to tribes, can have resounding effects around future access, use, and infrastructural regulation. It can take years to rebuild trust among potential partners. Tribes partnering with regional infrastructure teams must be clear about sovereign rights to taxation of infrastructure and services, obtaining rights-of-way, tribal environmental standards, conduct of non-tribal personnel, expectations around investment in workforce training, and other practices for solidifying sound business practices. States need to take into account the history of telecommunications build-outs and how this shapes the activity and business practices of smaller, regional service providers, especially in rural and reservation communities.

Another sovereign right critical to telecommunications and broadband deployment plans is the right to taxation. Non-tribal Internet service providers may expect to pay a state tax and a tribal tax for infrastructures and services crossing sovereign tribal lands. Recently the Law Office of Randal T. Evans and the Mobius Legal Group wrote a handbook of communications regulation and taxation in Indian Country after lawyers in both firms recognized the ways federal Indian law intersects with telecommunications taxation laws.[31] Much work remains to be done in this area. Indeed, part of Navajo Nation's reason for setting up the Navajo Nation Telecommunications Regulatory Commission is to clarify tribal taxation and licensing procedures for Native and non-Native service providers within Navajo Nation. Experience has shown that clarifying regulations for investors and entrepreneurs seeking to conduct business with tribes is a crucial first step toward encouraging competition through telecommunications regulation in Indian Country.

The tribal right to regulate the conduct and duty of federal employees also relates to local broadband deployment efforts and federal policy making and agenda setting. This is demonstrated through the development of the FCC's Office of Native Affairs and Policy, which occurred in large part because of the efforts of advocates of tribal media and the Clinton administration's 2000 Executive Order of Tribal Consultation.[32] Meetings between members of the Office of Native Affairs and Policy and tribal traditional historic preservation officers present a small-scale example of the exercise of this particular right, as the officers reminded federal employees of the need to create policies that require the recipients of subsidies to respect the laws around the protection of tribal sacred sites before constructing towers and stringing fiber-optic cable. In 2012, the FCC released a plan to begin meeting with tribal nations and intertribal governing associations to make sure that telecommunications towers comply with the National Historic Preservation Act.[33] Indeed, this research shows that tribes who build out their own broadband infrastructure gain political clout with various governmental and other institutional partners regarding telecom matters and the rights of tribes.

Undergirding all other tribal rights are the peoples' relationships within their homelands. This dimension of cultural sovereignty—deep knowledge of the landscape—applies to the decisions broadband project leaders make when deciding where and how to construct infrastructures. Broadband construction teams have to assess the lay of the land so as to identify optimal locations for setting up towers and dishes for line-of-sight wireless technologies, feasible locations for digging if laying terrestrial fiber-optic cable, and stable locations for stringing aerial fiber-optic cable. Tribes must cut new trails and roads for construction crews in addition to obtaining materials and hardware that can withstand extreme terrain and climates—rock-strewn canyons, hills, mesas, valleys, mountains, waterways, impenetrable forest, permafrost, high winds, monsoons, snow, and desert heat. This is why working with the traditional historic preservation officers, tribal archaeologists, and land management personnel is so important. Locations of sacred sites, plant life, ecological restoration areas, and migratory paths of wildlife must also be considered.

The history of colonial policies also influences treatment of the land in Indian Country, and this affects broadband deployments specifically with regard to obtaining rights-of-way for construction. Presently, many federal grant and loan criteria include a right-of-way requirement stipulating that construction teams acquire signed permissions from all landowners whose property will be crossed in the laying of fiber-optic cable or other necessary hardware. Unfortunately, obtaining these rights-of-way can be exceedingly difficult for tribes, in part because of issues related to tribal and state jurisdiction, unique treatment of customary tribal property inheritance rules, and property inheritance since the Dawes Act of 1887.[34] In some cases, tribes will want to build an infrastructure adjacent to state roads or state lands, but doing so requires obtaining rights-of-way to utilize cable that has already been laid under highways. In such situations, tribes must work closely with states, private landowners, or private business owners to gain permissions and, in some cases, to align their infrastructural plan with those of neighboring municipalities.[35] Other tribes must demonstrate to grant and loan program officers that customary tribal practices of property inheritance, such as matrilineal property inheritance, mean that there are no documented property surveys or assessments per se, but rather local oral history of land claims. Other tribes must contend with trying to obtain rights-of-way from multiple signatories who have inherited fractions of land since the Dawes Act, and who continue to reside on their family plots but do not have title to the land other than through genealogical evidence. Both the tribal right to regulate property inheritance and the cultural protocols around treatment of the land can conflict with US expectations of titled property claims, making it difficult for broadband teams to comply with rights-of-way requirements. Community members with deep knowledge of the tribal terrain can help broadband teams understand how the construction of a broadband network backbone interfaces with the geopolitical rhythms of the landscape.

It was with reference to the history of colonization and the ongoing negotiation of sovereignty that J. D. Williams, along with many other individuals working in tribal telecommunications, testified before the FCC in 1999, identifying tribal telecommunications and Internet service provision enterprises as examples of true self-determination.[36] At present, for the Native peoples of the United States, social and political power are crystallized in the form of tribal sovereignty. Within the flux of human relationships, the nature of power is such that it cannot be stored—it is not a given—but rather manifests through hundreds of smaller enactments across many interpersonal relationships.[37] Native peoples know that if they do not exercise the rights of sovereignty—legal-political and cultural—they will lose them to the US colonial imperative. This is why the first elder to speak at the 2010 NCAI Tribal Leader/Scholar Forum asserted the need for tribes to claim their sovereign right to airwaves. As Native peoples, we cannot expect that federal authorities will give these to us or hold them in reserve for us. Rather, we must assert the federal obligation in the context of earlier policies and court precedents and, strengthened by cultural sovereignty, make use of both the claims and the airwaves. Tribal leaders listened to this elder's important words. Advocates in the NCAI Telecommunications Forum issued resolutions on the matter. By 2013, the FCC Office of Native Affairs and Policy began investigating processes for creating a tribal priority for access to spectrum over tribal lands.[38]

This research has shown me that tribal command of broadband infrastructures and services undergirds the greater goal of helping Native peoples connect, communicate, and share information and knowledge critical to their survival after the century of disconnection enforced by early US industrialization and technological advance. It has also shown me that reliable and robust broadband Internet infrastructure and services in tribal lands will be integral to the exercise and ongoing negotiation of tribal sovereignty. Enforcing sovereignty means communicating with surrounding federal, state, and municipal authorities. It means building systems for transmitting key information quickly and efficiently. It means building systems in which to archive knowledge for future generations of tribal leaders. While now, in the contemporary historical moment, the rules of tribal sovereignty continue to be our most effective set of strategies for negotiating with federal authorities, younger generations of Indigenous thinkers and activists are imagining decolonial alternatives beyond the horizon of tribal sovereignty—new modes of governance that will circumvent colonizing hierarchies in tribal minds, lands, bodies, and ways of knowing. What do Native experiences with digital technologies—from mobile devices to network backbones—reveal about the functions of digital systems with regard to colonization and decolonization? We can ask ourselves these questions as tribal technicians, network administrators, policy advocates, entrepreneurs, and engineers continue to build out these systems across Indian Country. We are fortunate to have in our midst leaders with the know-how and political savvy to do so.

CHAPTER 7

Decolonizing the Technological

> Today it is only the professional who sees the imbalance and the general society comes to believe that the scientist can create the technology that is needed to bring balance back again. Thus, in spite of a clearly deteriorating physical world brought about by industrial society, we still think in mechanical, technological terms when we discuss restoration of what we have disrupted.
>
> —Vine Deloria, Jr., "Traditional Technologies"

DECOLONIZING METHODOLOGIES ARE DESIGNED TO CULTIVATE Indigenous and, more precisely, tribally centered solutions to community challenges. With regard to broadband Internet in Indian Country, specific research questions require further investigation for purposes of advancing network functionality and policy, in particular in low-resource (limited bandwidth, low income, challenging terrain, rural) locations.

Understanding the social impacts of large-scale digital technologies like network technologies also requires us to consider the multidimensional sensory, epistemic nature of digital systems and the ways Indigenous peoples live through their uses. We have to ask ourselves what makes a flourishing quality of life for Indigenous peoples in various places, given the current rhythms of globalization and the relationship of digital technologies to that quality of life. Pursuing such questions has us facing off against our conceptualizations of colonization and decolonization as manifested through our experiences with digital systems.

Thus this work requires continually reframing the intellectual stakes of research on Indigenous uses of digital technologies, in particular because research on technological phenomena is presently laden with assumptions about nationalist progress through techno-scientific advance. The goal is to acknowledge how industrialization and its continuation through the spread of ubiquitous computing and network technologies have shifted specific aspects of Native and Indigenous ways of life from previous particular eras. Yet, rather than presuming that mechanization and digitization will automatically improve Native ways of being, our goal as Native scientists and Indigenous thinkers is to push back at this false logic and understand the nature, goal, and direction of the solution that we are striving for when we do choose to apply digital technologies toward overcoming a particular obstacle in a tribal community. In a sense, careful investigation of the impacts of digital systems is about articulating the boundaries around these systems: their limitations, affordances, requirements, discrete effects, social contexts, and outcomes of their uses.

These kinds of investigations help us see and experience more distinctly when we, as Native and Indigenous peoples, are moving out of the reach of digital systems, and how our phenomenological experience of the world around us shifts, perhaps into a more embodied land-based or water-based Indigenous holistic manner of knowing, interacting, and sensing. Tribal communities will choose to apply digital systems as they see fit and will become accustomed to using that system in their own ways.

IMPROVING NETWORK FUNCTIONALITY

As tribes support more enterprises based on information and communication technologies, they are likewise encouraging the design of information systems, work practices, and devices for tribal uses. At the TribalNet 2013 conference in San Diego, I observed quite a few of the systems and devices designed to run on and support broadband connectivity in Indian Country, including wireless mesh technologies, transmitters and receivers, billing systems, data offload systems, database administration systems, and technical training operations. At the 2012 Tribal Telecom and Technology Summit, tribal liaisons with Sandia National Labs raised the issue of tribes acquiring and sponsoring top-level domain names, such as .hopi or .navajo, rather than the current nsn.gov designation. At the 2014 summit, John Curran, CEO of the American Registry for Internet Numbers, spoke about tribal domain names, as well as the importance of training technical staff to shift from Internet Protocol Version 4 to Internet Protocol Version 6 . Curran was quite clear: this is a change tribes need to make so that their systems continue to function properly.[1] At the 2015 summit, presenters from Sandia National Labs identified the need for tribes to begin developing cyber-security operations, in order to ensure the stability of their networks, including public safety networks. Other presenters spoke about tribal data centers and the economic impact these can have within the larger comprehensive economic development plan. There are few practitioners in the field, yet many tribal leaders are seeking to deploy and adopt network technologies while the field is rapidly changing, which makes research questions centered around the unique conditions shaping system and device functionality and uses in Indian Country all the more necessary.

Finding network engineers, computer scientists, designers, and data scientists who understand the rhythms of work and life in Indian Country continues to be a challenge, not only in terms of encouraging students to continue in this area of work through their formal schooling, but also in terms of finding faculty who can support these students and widen relevant scientific fields by publishing research in top journals and conference proceedings. Many systems scientists in academia come from a system-centered background, rather than a human-centered design and engineering background, and as a result often have little to no familiarity with social theory or the complexity of the so-called human factors shaping the design, uptake, and use of digital systems. As such, engineers and scientists who do take time to understand the social and political impacts of digital systems often span disciplinary boundaries in their work, explaining technical dimensions of systems and devices to social scientists and humanities researchers, on the one hand, and explaining humanistic and social concerns to technically minded researchers, on the other. This in itself represents a key site of intellectual labor when it comes to making space in which to challenge technocratic notions of progress, particularly with regard to the rights and experiences of Indigenous peoples.

INTERNET AND TELECOMMUNICATIONS POLICY IN INDIAN COUNTRY

Broadband policy continues to be of prime importance as each of the federal agencies charged with deploying infrastructure through Indian Country complies with the tribal consultation order. The constraints of working in telecommunications in Indian Country—particularly around developing tribal enterprise, taxation, acquiring spectrum licensing and rights-of-way, federal subsidy programming, and protecting tribal members' data, privacy, and intellectual property—will continue to be debated in Congress in the years to come. In 2015, representatives from the Government Accountability Office began an investigation into these issues. After interviewing representatives from twenty-one tribes, attending intertribal assemblies about tribal Internet and telecommunications issues, and visiting with tribal officials and managers of Internet service providers, representatives from the Government Accountability Office (GAO) presented their findings before Congress. Major findings revealed a telecommunications field dominated by private industry interests and highlighted the need for increased coordination between federal agencies supporting Internet and telecommunications access, subsidies to promote Internet access for tribal schools and libraries, and development of the means of collecting data about broadband connectivity and coverage in Indian Country.[2] While the GAO report is intended to inform policy makers, from a network science perspective, compelling research questions rise out of the need for more specific and accurate broadband coverage data in Indian Country. Who should gather these data, and how? How can the information be verified? How might this kind of data, visually expressed in the National Broadband Map, shape the nature of discussions around access, uses, coverage, and connectivity? What scientific advances, such as through data compression, network engineering, and wireless mesh interventions, can we help design in Indian Country for tribal communities? How can we prepare the data so that tribal decision makers can view and understand how this information affects their communities? At this point, I suspect that the coverage and access data will contain evidence of Internet service providers inflating coverage and subscription rates to Native communities in order to acquire federal subsidies that will go toward improving services to other locations.

Additionally, working in this field means learning more about the ways innovation in Indian Country occurs through the nexus between private industry, tribal government, tribal enterprise, federal government, universities, and intertribal policy-making organizations. It also requires learning more about the rhetoric and diplomacy that shape information and technology policy efforts in Indian Country, and how these relate to research projects emerging from the scientific community, the tribal law and governance communities, media and representation communities, and economic development communities. Each of these fields of work comes with different stakes and goals, and it is a challenge from the policy-making perspective to synthesize divergent and convergent findings into compelling and actionable policy narratives. There are various issue networks surrounding large-scale digital infrastructure, each with its own goals and objectives. From a social scientific perspective, one can acknowledge that large-scale digital infrastructures—and especially networks—are in many ways epistemically multifaceted, but from a policy perspective, this translates into competing goals and rhetorics across lobbyist groups, think tanks, industry partners, congressional representatives, and tribal interest groups. The way one frames narratives about data, information, access, and digital technologies in Indian Country shapes the deployment of policy and regulation at multiple stages and levels of governance, from community decisions to allow WiFi access in tribal libraries to federal decisions about sovereign tribal rights to spectrum.

DIGITAL TECHNOLOGIES AND INDIGENOUS QUALITY OF LIFE

Conversations about the impacts of Internet and digital technologies in Indian Country inevitably lead to discussions about how we imagine and define quality of life in Native communities. With regard to theory, the work of understanding the dimensions of the Native cybersphere continues to be of interest. What devices and systems undergird the design of this Web topology removed from the physical environment yet deeply ingrained with messages about homelands and sovereignty movements? How do Native and Indigenous peoples utilize an array of broadband technologies toward the furtherance of explicitly Native and Indigenous goals? What might this contribute to our knowledge about the ways marginalized social groups organize politically and socially and mobilize via available digital systems and devices, particularly across locations where Internet access is rare or unaffordable? How are spectrum, the World Wide Web, the Internet, smartphones, unmanned aerial devices, and other such systems theorized and experienced within specific Native ways of knowing?

More than once in the course of this research, I heard tribal peoples, especially of the southwest, refer to the Hopi prophecy of Spider Woman weaving a web around the world, indicating an era when the original peoples would come back together.[3] I am not Hopi. It is not my place to undergo the intense linguistic, spiritual, physical, and philosophical training needed to fully understand this prophecy. Yet many times I have observed the conversation shift when this prophecy is mentioned during discussions of intertribal adoption of broadband systems and devices. There is a spiritual and philosophical dimension to the work of deploying and designing networked systems throughout Indian Country. Finding ways of teaching pragmatic social scientists to hear and make sense of this level of data—sacred landscape and sovereignty, or place and spirituality—as they consider the particularities of information system design, technological infrastructure, policy making, and information flows in Indigenous communities is central to this work, both in theory and in practice.

More deeply than that, there is the work we have to do in opening the door for Native youth who are born into ways of life that are deeply enmeshed within the cybersphere. Traditionalists continue to evoke the profound spiritual and philosophical significance of subsistence hunting, prayer, traditional foods practices, and teaching youth to shape clay, grasses, bark, and stone for the continuation of Indigenous arts. Younger generations of Native artists, activists, students, designers, engineers, and entrepreneurs appreciate traditional practices at the same time that they utilize digital tools and techniques for expressing aesthetic visions, organizing protests, learning and teaching, designing buildings, power grids, reservoirs, and health delivery systems, and supporting Native-owned businesses. Digital tools and techniques are not necessarily incompatible with traditional tribal ways of life. Rather, younger generations move in and out of the cybersphere as seamlessly as they move onto and off reservations, through densely populated cities and solitary mesas, from downtown burger joints where even paper cups come at a price to potlatch gatherings where giveaways are the norm.

On the one hand, it is important to accommodate the skills and aptitudes of the digital generation of Native youth. On the other hand, it is likewise important to keep teaching future generations about the limitations of digital devices and methods and remind them of the grace and peace that can be achieved through traditional non-digital practices. Thus tribal governments must invest in supporting digital tools, techniques, and systems with the aim of strengthening administrative efficacy and tribal member employability even as they also invest in traditional arts and language lessons. While digital devices such as smartphones give us the impression that we can speak with anyone at any time, and media apps attune us to particular fast-paced norms of play and interaction, traditional practices such as firing clay pottery, carving canoes, weaving baskets, and observing long, sacred dances also remind us about the rhythms of community engagement, the order of seasons, patience in making tools that are pleasing to the eye and that work well, and use of the physical senses beyond video-game twitch speed and the rapid eye motion needed for scanning flashing digital images.

Thus we have to ask ourselves about the nature of Native and Indigenous approaches to quality of life, and how these are shaped through reliance on digital tools and techniques. How do we teach young people to value the smell, feel, and purpose of the grasses, woods, water, and earth of their tribal environment, while also teaching them to appreciate digital connectivity? Both these ideas are rooted in conscientious orientation to relationality. The idea that all beings, phenomena, and ideas are deeply interwoven and interconnected is foundational to Native science. It is no matter of happenstance that references to Spider Woman weaving a web around the world arise in many Native conversations about digital technologies. Rather, this powerful story and others like it indicate both Indigenous philosophical explanations of digital technological phenomena in Native communities and a Native scientific paradigm. As the study of ecological and environmental sciences lends itself to a deeper understanding of this concept, so do studies of ICTs, insofar as thinkers in these fields embrace human-centered, community-centered approaches and position their work to make environmentally healthy interventions in philosophies of science.

Above all, it is not possible to have a conversation about quality-of-life matters for Indigenous people without understanding colonization and theorizing decolonization. The concerted effort of many Native and Indigenous activists, elders, scholars, students, artists, policy experts, and leaders speaking with one another over time has resulted in resistance projects with philosophical and moral goals that contribute to what we now recognize as Indigenous decolonization. While there are many expressions of decolonization and decoloniality, Native and Indigenous studies scholars usually base their pathways to decolonization on the experience of Native peoples who have endured Western European and Euro-American colonialism. To be clear, speaking about decolonization from an Indigenous perspective means not only interrogating racist, sexist, and oppressive colonial structures and logics; it also requires creating the conditions for the total divestiture of the occupying foreign powers in Indigenous territories.[4] Conscientious conversations about decolonization point to an unforeseen era in which the United States, Mexico, and Canada fundamentally shift their sovereign government-to-government relationship with sovereign and autonomous tribal governments. It indicates the end of the domestic dependent status of federally recognized tribes in the United States and the beginning of a new mode of diplomacy, adjudication, and order. It requires that Indigenous thinkers and tribal leaders carefully consider the limitations and needs of their communities and what values, discourses, epistemologies, rules of order, and modes of governance can be established so as to provide for the community as it is in the contemporary moment, carrying

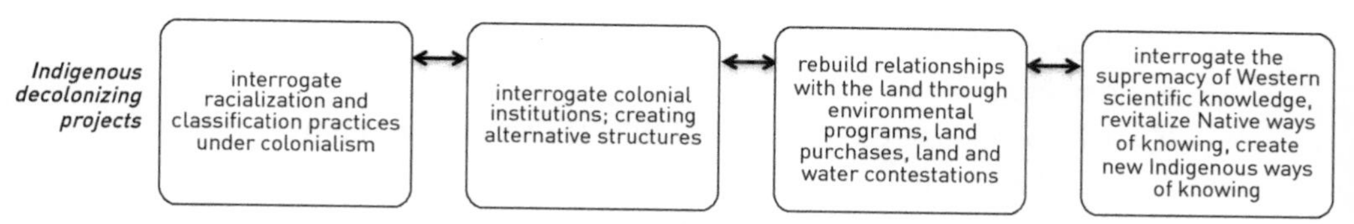

FIGURE 7.1. At present many Indigenous peoples around the world are undertaking various kinds of decolonizing projects. Read against the coloniality of power, many of these decolonizing projects are often interlaced and focused on (1) interrogating racist, sexist, and classist colonial classification practices; (2) interrogating colonial institutions and creating alternative Indigenous institutions; (3) rebuilding relationships with the land and waterways through restoration and claims processes; and (4) interrogating the supremacy of Western scientific knowledge by revitalizing Native ways of knowing.

all the traumas of past colonial eras, including the scars of poverty and internalized oppression, willful ignorance at times, and at other times profound spiritual commitment to overcoming the despair of reservation life.[5]

Initiated largely as knowledge projects, decolonization in the context of Native North America is often focused on (1) reclaiming tribal and Indigenous understandings of the self with relation to settlers, people of color, and other tribal people, as well as reviving Indigenous sensibilities around belonging and community responsibility; (2) revitalizing non-European Native and Indigenous languages, philosophies, practices, narratives, and spiritualities; and (3) restoring Indigenous biomes, including lands and waterways, and the ways of being that cultivate Native relationships with landscapes. Decolonization moves beyond anticolonial critique and includes building Indigenous programs and institutions that support self-determination and self-governance. Figure 7.1 shows how these reinforce one another.

As Eve Tuck and Wayne Yang assert, decolonization is not a metaphor, nor is it a euphemism for a way of life free of the settler imperative to conquer and assert white supremacy. It is about a new independent state of governance in which Native and Indigenous peoples assert their rights in spite of the settlers' imperative to colonize. Considered to its full extent, decolonization is a world-historical idea, a set of concepts and theories derived from various experiences within the third world, Eastern bloc countries, black, people of color, Latin American, and Indigenous communities.[6] It is helpful to acknowledge the complexity of the idea; it is a multilayered wicked problem for Indigenous scholars. Like technology, it has the capacity to operate like a black box. Neophytes use the word lightly and apply it to any modality that is antiracist, antisexist, anti-neoliberal, or anticolonial, with little regard for the vastness of the idea: a status of Indigenous independence, the return of tribal lands, or, as Leslie Marmon Silko wrote in 1994, a total spiritual reclamation of the homelands.[7]

If we acknowledge the vastness of the idea and make room for it in our minds, after having learned about the challenges of tribes acquiring access to the Internet—a globally networked series of infrastructures built around a set of standards—we must then return to the concepts of human interconnectedness, the technical interoperability of this critical communications infrastructure, and the matter of diplomacy and regulation. Through relationality, Indigenous thinkers are constantly connecting ideas, beings, emotions, and words in the mind, heart, spirit, and body, making decisions about what to foreground and what to background with regard to how one sees and interacts with others. What aspects of the methods, techniques, and materials of the eras of industrialization and technological advance—technocracy—have we already deeply internalized? How have we learned to sustain Indigenous political subjectivities, spiritualities, and philosophies alongside our own technoscientific rationales and industrialized digital habitus? What kinds of political, ethical, moral and spiritual, social, and scientific discussions should we be having around the relationship between tribal governmental command of digital technologies, uses of ICTs by Native peoples, and Indigenous pathways to decolonization?

Understanding how Native and Indigenous peoples use analog and digital systems to share knowledge toward self-governance and self-determination—by talking story, by sharing information and data with one another—gives us insight into the subversive ways that Native and Indigenous peoples apply digital technologies toward creating spaces for Native and Indigenous forms of resistance, endurance, and liberation. Many times I have sat with friends and imagined how the Native world might be if tribes were digitally well connected: artists collaborating with artists, tribal leaders accessing databases before writing policy, health practitioners consulting patient records at a distance and completing digital health checks on elders high up on the mesas or across the lakes in wild rice country. We have imagined e-commerce, e-government, and digital networking. We have imagined massive digital libraries, digital language learning, and public safety networks. Figure 7.2 shows the relationship between Indigenous praxis (the actions that Native and Indigenous peoples engage in order to subvert colonization) and the function of systems (databases, social media networks, data-sharing agreements, reporting structures, mapping, storywork, gaming, library and archival projects, digital language-learning tools, etc.) that in combination give us the intellectual

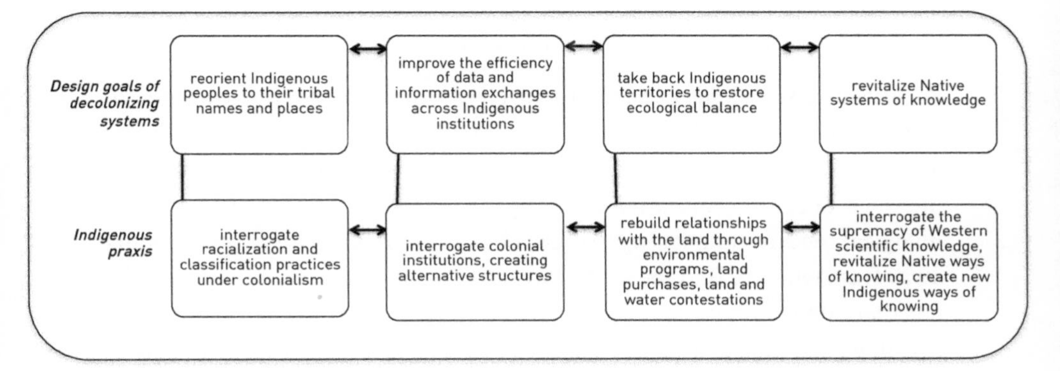

FIGURE 7.2. Native Americans are increasingly designing information systems to support Indigenous decolonization goals and everyday praxis. Many of these systems are designed to (1) reorient Indigenous peoples to their tribal names and place-names; (2) improve the dissemination of high-quality data, information, and knowledge across tribal and Indigenous institutions; (3) support tribal land claims, environmental management, and ecological restoration; and (4) revitalize Native ways of knowing among younger generations of Native American and Indigenous peoples.

space to consider Native and Indigenous peoples' command of data, information, and ways of knowing toward the divestment of colonial authority over Indigenous ways of life.

While it is technically possible to carry out various projects of this kind, the digital interfaces that many of us create are not going to restore the way of life that existed before the scourges of the colonial dynamic. Those worlds belong to a different time and a different people—our ancestors—and while, as Indigenous thinkers, we still speak tribal languages and practice our spiritualities and philosophies, we are also continuously adding new technical terms to our vocabularies, widening our political discourses, and reinforcing Indigenous ways of knowing by collaborating across a digital episteme.[8] The interfaces, devices, and systems we build are tools that help us work together and collaborate toward creating a way of life that is very much shaped by the dynamics and needs of the present and, with vision and heart, can help us create the conditions for a healthier future. The technology can take us only as far as we can see. In that sense, the construction of sociotechnical systems can also be limiting, and when we take that seriously, when we incorporate that into our planning and creativity, it is easier to step away from design thinking that grounds engineering work, the political strategizing that grounds policy work, and instead move toward understanding the long-standing truths embedded in Native storywork.

REFRAMING TOWARD INTELLECTUAL DECOLONIZATION

Sociotechnical systems in and of themselves will not create the healthier visions that move beyond colonial governmentality. Those long-standing visions and insights emerge from an episteme that is not dependent on the platform, interface, or device. For Native and Indigenous peoples, decolonizing the technological requires unpacking the black box of technology and then stepping away from what we find there. We have to take ourselves out of that box: the box that says we are premodern and therefore antitechnological, the box that says that Native peoples do not do well in STEM (science, technology, engineering, and math) fields, the box that says that an investment in digital technologies is robbing us of our most sacred traditions, and the box that says that digital technologies are the best solution for the "Indian problem." We have to challenge and interrogate those false logics. Then we must learn how to create something better: a new scientific subfield, a more appropriate design space, a more stable enterprise, a more knowledgeable tribal government, labs with better instruments, and thinkers with renewed respect for what Vine Deloria, Jr., referred to as traditional technologies, the holistic ways of knowing that precede physical instantiation.[9]

At this point, it is important to acknowledge not only the dearth of scholarly literature on the intersection of ICTs with the lives of Native and Indigenous peoples but also the degree and nature of the inaccuracies in the literature that is available. A good amount of inaccuracy in the scholarly depiction of Native peoples' uses of ICTs emerges out of the prevalence of colonial logics at play in many social scientific fields of study. Researchers in the broad field of information science are not necessarily trained to account for the realities of colonialism or, more specifically, the sovereignty and autonomy movements and legal frameworks that underpin Indigenous experience. Similarly, researchers in the broad field of science, technology, and society studies, while able to engage social scientific accounts of power to explain social difference, have contributed few studies dealing with the particular forces inscribing US Indigenous experiences. Media studies researchers—communications, journalism, arts and aesthetics, film and media, and cultural studies—have descriptive studies on topics such as film and media in Native American communities, community radio initiatives, and quite a lot on the topic of media representation of Native peoples in their archives, yet the disciplinary orientation of media studies precludes the pragmatic engineering orientation—emphasis on design and behavioral outcomes—that guides much of the inquiry in technical scientific approaches. Thus there is much description of possible media effects and the matter of gaze but less on the functionality of information and devices as habitus in Native peoples' lives.

There is one body of work that effectively sidesteps these challenges: the research and scholarly articles arising out of informatics researchers' and legal scholars' engagement with the K-Net project in northern Ontario. Brian Beaton, Susan O'Donnell, Michael Gurstein, Richard McMahon, and their colleagues are among the first to bridge information scientific methods and discourses with an understanding of the policies and histories shaping uses of ICTs within First Nations reserves.

Overall, however, few scientists have taken such an approach with regard to the unique policies and histories shaping uses of ICTs with regard to Native Americans and legacies of US colonialism. As Native and Indigenous scholars well know, the social policies employed by the authorities of modern nation-states to deal with their indigenous "subjects" very much shape the ways Native and Indigenous peoples generate corresponding modes of governance toward community resilience and survival. To the untrained eye, aboriginals, Indians, American Indians, Native Americans, and indigenous peoples are all marginalized ethnic minorities in more perfect, modernized, technologically advancing nation-states. The temptation is to compare all Indigenous cases regardless of time and place, and worse, with disregard for diverse Indigenous geopolitical contexts and distinct philosophies, histories, languages, and spiritualities. US policies shaping ICT infrastructural access across American Indian reservations must differ from Canadian policies devised for the same purpose but implemented across First Nations reserves. Furthermore, the people governing each unique tribe will deploy, adopt, and apply ICTs in the manner most beneficial for the continuation of their cultural sovereignty within their tribal community. While the very functioning of the nation-state depends on categorical standards around division, social ordering, and centralized rules of command, the very nature of Indigenous peoples' self-governance is based on rhythms and ecologies within local landscapes. Tribal governance is aimed not at expansion but rather at sustaining specific lifeways. Understanding this distinction requires understanding how sociotechnical infrastructures are nested within historical sets of values and relationships that are not necessarily oriented toward ideologies of progress, expansion, assimilation, the conquest of nature, and the supremacy of technological advance.

For this particular investigation, one of the most difficult aspects, methodologically, of entering into this line of inquiry had to do with first identifying those causal logics—in this case, forces of colonialism and sovereignty—grounding social and political moves in Indian Country and, second, structuring analytical frameworks for making sense of ICTs within that context. Reframing allowed me to bridge what the field of Native and Indigenous studies has to say about colonialism with what the field of information science has to say about information and devices in daily life. Doing so consisted of positioning myself, both physically and epistemically, in particular places, in the company of particular thinkers, and within a particular orientation toward the literature and application of information science.

The first assumption I had to challenge was the social scientific researcher's fear of "going native," a racist phrase and concept stemming from the early anthropological practice of living among Natives for purposes of ethnography yet rejecting the perceived Native customs of interacting, dressing, and thinking, for fear that white adaptation of the Native manner of relating in the world might skew the white researcher's objective assessment of Native culture. Veiled in the language of science, the threat of "going native" is still discussed in qualitative methods texts as a concept indicating a relational distance that a researcher must maintain from the company of human subjects.

As a Yaqui woman who is also a researcher and scientist, I find the concept of "going native" troubling and illogical, both for its denial of the power dynamic between the researcher and the researched and for the tacit assumption that not only do Native peoples have nothing to teach scientists but that their very manner of relating with the world somehow negates the credibility of scientific research methods. (Imagine the reaction in the social scientific community if an elite set of Indigenous editors were to support the systematic publication of qualitative methods chapters on maintaining theoretical credibility by avoiding "going white.") In short, I entered the field of information science with a keen awareness of positionality. I see information science as a twentieth-century applied science, and as such, the foundation of its theory making, practice, and field building is bound up in the twentieth-century global circulation of goods, labor, and currencies that depend on the continual exploitation of Native lands, waters, and bodies. The gift of my people's histories and spirituality helped me perceive a dimension that earlier studies of Native uses of ICTs failed to inscribe and that relates to the inherent sovereignty of Native peoples choosing to use ICTs and build the infrastructure for it across their sacred lands toward their own tribal goals, even when non-tribal critics, including Indigenous scholars, decry such efforts as a perpetuation of corporate colonialism, neoliberalism, technological hegemony, and other such challenging allegations.

I was also able to discern the difference between what Lakota scholar and architect Craig Howe refers to as design for Native nationalist purposes and design for tribal purposes.[10] What the Indigenous Information Research Group operationalized as Native ways of knowing—those ways of perceiving, sharing, and creating data, information, and knowledge that are founded in unique and distinctly non-European tribal epistemologies—maps over to what Howe describes as tribalism. In 1998, he asserted that "there is no place for tribalism in the cybersphere," and more than a decade later, nested in the wired and digitally charged city of Seattle, I could agree.[11] Building out broadband networks shapes tribal governance work and economic development efforts. It is a part of Native nation building, but it is not essential to the maintenance of unique and distinct tribal epistemologies. In Hopi, the ceremonial people pray and the rains fall regardless of the mobile devices in pockets and purses. In the Pacific Northwest, the canoe families spiritually and philosophically nourish themselves and whole communities while paddling in rhythm with the waves and with one another regardless of the availability of a cellular signal. Sundance does not need a laptop. As Subcomandante Marcos writes, "The issue is not about what you write with, but the hand that dreams when it writes. And that is what the pencil is afraid of, to realize that it is not necessary."[12] But I could pick up on this kind of distinction—between web epistemologies and Indigenous paradigms—only because my family members had taught me when and how to see with Indigenous eyes. I bore an Indigenous intellectual lineage before I came to the university.

Once at the university, I could identify with feminist thought and, through that, trace my way to the concept of positionality. Frequently employed in critical race theory and gender studies, the methodology of positionality requires researchers to identify their own degrees of privilege through factors of race, class, educational attainment, income, ability, gender, and citizenship, among others, before seeking the epistemological basis of their intellectual craft. Doing so helps them understand how their way of making meaning, of framing research, within their conceptual universe is tied to their positionality within an unjust world.

Though reframing is conceptually related to the methodology of positionality, it is more specific in orientation, as its precise goal is intellectual decolonization through the correction of white settler supremacist explanations of a social problem or challenge within an Indigenous community. Before researchers can reframe a social problem and diagnose an intervention, they must see themselves and their conceptual universe in relation to the nature of the problem and, from that point, make decisions about what to foreground in the assessment and depiction of the problem. Through my first phase of fieldwork, I learned that I had to reorient my discourse, values, and thinking away from academic information science and toward the values of Indian Country. The people to whom I spoke in Indian Country did not talk about devices, design, information systems, information flows, or information as power. They talked about creating jobs for tribal youth, respecting the stories of the elders, complying with the White House tribal consultation order, and driving all-terrain vehicles up sacred mountains in search of safe places for building towers for their people. I had to think about how to use the methods and concepts of information science as tools for advancing intertribal goals in Indian Country. Only by doing so was I able to perceive how Native moves toward sovereignty align with Native uses of ICTs. This is what makes reframing a decolonizing methodology: it reorients the techniques of applied science toward meeting the goals of tribal communities. That reorientation reframes the nature of the challenge. In this case, I shifted the inquiry away from "Why don't Indians have access to the Internet?" to "In what ways do tribal leaders acquire Internet for their communities, and what can we learn from their approaches?" I admit that when I entered the field, I had to unlearn the legitimated academic discourse so that I could listen more clearly for the voices and experiences of individuals sharing narratives rooted in social subjugation. My research "subjects" taught me; they allowed me to participate in what they had learned.

Thus this work seeks to reveal what undergirds the phenomena in question—tribal uses of ICTs—as well as what undergirds the existing social scientific explanations of these phenomena. The tribal command of affordable and robust broadband infrastructures across reservation lands undergirds tribal peoples' uses of ICTs and the development of corresponding information systems and social programs. Applying an Indigenous orientation to the review of the literature and conceptual framework reveals the colonial logics at play in earlier studies and opens the field to alternative explanations rooted in the experience of the Native peoples who are building these systems and devices. This investigation of tribal uses of ICTs reveals (1) how alternative histories and governmental paradigms are embedded in the design values and uses of ICTs by Native peoples, (2) how the colonial conditions shaping Indian Country in turn shape access to information, systems, and devices, and (3) how issues of access are deeply political and historical. Thus, while the social and economic policies of a dominant governmental hegemony very much constrain access and uses of ICTs by that government's marginalized citizenry, the intentions and goals of sociotechnical system designers imbue these systems with a set of possibilities that reach through and beyond the dominant government's technological discourses.

This means that if, as scientists, we are to learn about the social dimensions of access to information and ICTs, we must be able to explain how globally elite classes of technologically advanced government and industry leaders sustain social and economic policies that productively marginalize particular vulnerable populations. Poverty, lack of infrastructure, distinct cultural values, physical geography, and other commonly cited explanations do not cause the lack of access to ICTs in marginalized communities but are, rather, symptoms and conditions of long-term social policies that depend on limiting flows of data, information, and knowledge within these communities. In this research, I identified the forces of colonialism and tribal moves toward sovereignty as the logics shaping Native peoples' efforts to acquire broadband infrastructures across reservation lands. Marginalized peoples all over the world are not marginalized by ghosts and paper; the active leaders of governmental regimes write and enforce oppressive policies every day. Thus, understanding uses of ICTs in marginalized communities requires understanding the roots of policies affecting the modes of oppression operating within that community every day.

COMMUNITY OBJECTIVES, COMMUNITY CALIBRATION

The study of digital technologies is certainly one of the many areas that span academia, industry, and policy in which it is all too easy to forget the integral experiences of specific communities of use. One of the many assumptions shaping the myth of progress through techno-scientific advance is that of the homogeneous digital user, the inevitable consumer in every place in every country who desires to partake in Thomas Friedman's globalized "flat" world.[13] While there are certainly global technical standards regulating Internet use—the interoperability of computing languages and designations of file types, for example—there are nevertheless various imaginaries shaping Internet use in specific places.

In Native communities, conversations about the limitations and utility of digital technologies happen in ordinary settings as individuals choose to employ various digital technologies to resolve everyday problems. A 2010 study of the commentary around Pueblo people posting videos of sacred dances on YouTube revealed the nature of these community discussions.[14] While the general rule in many Native communities is to avoid recording prayer, sacred practices, or moments of reverie—indeed, many tribal governments and spiritual societies have strict rules against it—the discussions in the commentary revealed careful consideration of special exceptions, such as when relatives far from home due to military service, work, or schooling are in need of the healing afforded through witnessing a dance. The deeper meaning of such conversations points to the ways Native people embed tribally specific spiritual and social intentions in their development of social media posts and networks.

In a 2013 interview, Seattle-area Idle No More activist Sweetwater Nannauck described this dimension regarding her strategies for disseminating communications about upcoming flash mobs. When asked how she decides to organize protest rallies, she answered that they are prayer rallies, not protest rallies. The intention to organize begins with prayer, guidance from ancestors, and renewed commitment to caring for the homelands.[15] A flash mob presumes that an array of strangers connected through multiple social media platforms, including basic text messaging, are willing to act in unison based on a series of pushed digital commands. While the flash mob is thoroughly a product of willingness to participate within a digitally connected social order, the way that Idle No More flash mobs coalesced in the winter of 2012–13 also depended on the participants' commitment to a thoroughly Native American and First Nations social and philosophical order. Idle No More flash mob participants self-organized on the fly with hand drums, sage, Native rights activist T-shirts and sweatshirts, full powwow regalia, Northwest coast button blankets, and carefully woven cedar hats and headbands and brought songs they were willing to share. They were willing to be recorded. They wanted digital images of their presence to circulate through various media channels so that many people would hear their message. They drummed and prayed at malls, border crossings, parks, and other public places, intentionally transforming the very spatiality of an anti-Indigenous neoliberal social order through claiming the relationship with the earth, waters, plant life, and air, all the features of Indigenous homelands that a neoliberal technocracy seeks to exploit and capitalize.

Vine Deloria, Jr., writes about Native people's traditional technology, positioning Native youth's attunement to ecological harmony in opposition to the colonial technocratic citizen's dependency on postindustrial technological advancement.[16] He criticizes the habitus of the technologically advanced colonial authority, who self-assuredly diagnoses all the problems in Indian Country as the result of lack of adequate technology and then pays himself or herself handsomely to repair the Indian problems with government-funded industrial solutions. Colonial governments seeking the assimilation of Native peoples provide training and education for Native youth, who, forgetting their commitment to the ecological harmony of their homelands, develop superficial solutions to so-called Indian problems. Native peoples' concern about assimilation through participation in a neoliberal technocracy is significant. This assimilation is possible and bears very real consequences for tribal groups interested in decolonization and the restoration of tribal governments based on tribal philosophies and leadership practices rather than settler-colonial forms of techno-scientific advance and government by surveillance. For this reason, part of our work as Indigenous scholars is to identify the ways colonization and co-optation work and then to communicate what we learn, in particular with our tribal communities and leaders. Lessons from Third World decolonization efforts are rife with examples of hopeful Indigenous elite who end up fomenting new kinds of colonization, channeling strategies that center the neoliberal nation-state in tribal affairs rather than centering tribal self-governance based on the needs of the local community.[17]

This is not to say that all digital technologies, or any technology systems that are the fruit of the Industrial Revolution, are inherently colonial with regard to the contemporary status of Native and Indigenous peoples. Rather, it highlights questions of power, directionality, design values, and the economic privileging of particular technological systems and infrastructures that serve the reification of colonial authority. It is possible to design digital technological systems that serve Native peoples in their efforts at cultural revitalization and the strengthening of tribal governance. The members of AbTec, the Montreal-based Aboriginal Territories in Cyberspace collective, engage these questions through development of digital gaming environments, Indigenous Machinima, and activist blogs designed to create a home for Native peoples in the cybersphere.[18] Many of these projects continue to focus on the decolonizing work of asserting tribal histories, centering specific Indigenous aesthetic approaches, and creating platforms for reinforcing lessons about traditional foods and medicinal practices. Games like NeverAlone/Kisima Innitchuna, produced through a partnership between Iñupiaq storytellers and artists and game designers from Upper One Games and Playstation, offer rich grounds for understanding the kinds of learning that takes place for Indigenous peoples through digital environments designed for play.[19] Understanding the way Native and Indigenous activist groups around the world connect with one another and form networks of solidarity through platforms such as Twitter and Facebook may also yield evidence of the reach of Indigenous political messaging and

Indigenous anticolonial expressions as they arise in locations around the world. These kinds of digital environments represent the application layer of global Internet infrastructure.

While applications—apps, games, software, user interfaces—provide rich grounds for understanding Native peoples' relationships within particular digital environments, the preceding chapters in this book point to the need to understand with specificity the power that Native tribal leaders have in the continuing political, technical, and social innovation of the Internet as regulated infrastructure. In the spring of 2015, the island nation of Nauru, with a population composed mostly of Indigenous islanders and detainees held in a prison sponsored by the Australian government, blocked all access to Facebook.[20] Select Internet sites were also blocked, as prisoners' rights activists and Indigenous rights activists had been utilizing social media to spread awareness of human rights abuses on the island. Native and Indigenous peoples around the world have a great deal of experience with the very real power of select governments that shut down communications by blocking access to popular forums. We experience it when Native histories are precluded from public school coursework. Previous generations still suffer from the dehumanizing punishments of childhood, when missionary teachers abused them for speaking their languages. Native independent media journalists continue to seek stable outlets for disseminating news about critical issues throughout Indian Country. That tribes in the United States and Canada are increasingly taking command of the build-out and policy making related to communications infrastructure on, around, and through their reservations is a very powerful and welcome move, in particular as colonial warfare now cycles through cybernetic and informatic means.

Native youth, who are increasingly comfortable living through the cybersphere, face the challenge of figuring out the dimensions of power and struggle that emanate from digital technologies, how these relate to the colonial present and also evoke a range of possibilities for decolonial futures. This work requires methodical dedication. The task of harnessing digital technologies to shape Indigenous futures is a project of its own within Indian Country. The goal is not to denigrate responsiveness to the natural world or supplant Native ways of knowing with desires to embody the digital in social and governmental forms. Instead, younger generations are applying technological tools toward retaining values, histories, and philosophies attuned to the sacred communal ecologies and rhythms that make us Indigenous.

Conclusion

NATIVE ARTISTS AND INTELLECTUALS ARE INCREASINGLY CONSIDERing the ontological stakes as Native peoples integrate digital technologies into their everyday lives. While considering topics such as Indigenous posthumanism and the relationship between artificial intelligence and Indigenous knowledge is a remarkable philosophical exercise, it is important to focus first on the material aspects of the Internet—fiber-optic cables, server rooms, towers, wide area networks, and Internet service providers—and how these come to be on Native land. Understanding the tangible material aspects of the Internet helps ground philosophical discussions about digital technology and reminds us of the formidable labor required to provide 24/7 high-speed Internet access to rural, low-income communities and underserved and historically marginalized populations. It reminds us of the market forces shaping the privatized Internet and telecommunications industry in the United States and helps us consider the economic impacts of tribes as governments investing in these costly yet vital communications infrastructures.

Considered against the backdrop of Native nation-building efforts, decolonization efforts, and policies of self-determination—all of which point in different directions when it comes to Native American and Indigenous resurgence throughout the United States—network backbones inspire compelling visions about the potential for digital technologies in Indian Country. It is while we are imagining those visions—talking about them, investing in them, designing pilot projects and start-ups, creating new aesthetic practices, wiring our government buildings and hosting web pages—that we must pay attention to what we are experiencing and thinking as we weave digital practices as Native peoples into our lives. Digital technology projects function in some sense like a mirror, reflecting back at us what we expect them to help us overcome, with the systems we design revealing our own methods for classifying, categorizing, and making sense of data and information. Through our uses of digital systems as Indigenous peoples, they become embedded with what we believe to be our Indigenous values. Because of their scale, broadband network backbones in Indian Country are epistemically multifaceted and are built through the instantiation of the values of the many individuals, teams, and institutions that support their construction and improvement.

Acknowledging those values helps us as Indigenous peoples to check what we believe these systems are doing for us against the kind of labor that, as Latour might say, they are actually entraining us to engage in. Thinking about values and practices around the design and deployment of large-scale infrastructures leads us to consider the will of the communities who build their own infrastructures. Martin Heidegger and Vine Deloria, Jr., share this concern: they both wrote about how exercises of power pertain to human will and, more specifically, when it comes to governance, to the will of a people.[1] Native peoples know that colonization programs are not realized through the will of an individual but rather require the sanction of state governments and authorized institutions, as well as their associated personnel. Tribal leaders know that sovereignty is ongoing work and that if, as inherently sovereign Native peoples, we do not exercise our legal and political rights—enforcing treaty obligations—we face the threat of US settler colonial encroachment on Native lands and waters. Information scientists conceptualize how individuals use information and communication devices as extensions of themselves. Heidegger wrote about the threat of technologically advanced people seeing the technological advance itself as their purpose for being and forsaking it as the means toward more morally upright purposes.[2]

Here is where Heidegger and Deloria part ways: Heidegger sees man as a technological being separate from the natural world, while Deloria sees humans as deeply in rhythm with a constantly unfolding cosmic natural order. When it comes to individuals using ICTs en masse in order to self-govern and politically mobilize, Heidegger views ICTs as a large-scale means for dominating nature and, in so doing, fashioning humanity toward a superior, though deeply flawed, state of technological advance. Deloria reminds us that as Native peoples, our obligations in utilizing technologies are to strengthen our relationships to our homelands and to cultivate wisdom from the patterns we experience there.[3] These two philosophical approaches to technology indicate distinctive, at times divergent, and contrapuntal trajectories regarding the social impacts of large-scale adoption of network technologies. The human orientation toward domination is central to both approaches. We have to acknowledge that Heidegger was writing during the rise of the National Socialist, or Nazi, Party. His intellectual life was very much shaped by the politics surrounding Nazism, the country's economic fears, and the effects of German nationalism not only in Europe but also on the world geopolitical stage. We also have to acknowledge that Deloria was writing a generation later, during the Vietnam War and the interventions of COINTELPRO in the American Indian Movement, during an era of American Indian revitalization and resurgence, an era that continues to shape our present. When we think about when and where these two approaches emerged, we can appreciate the nuanced relationship between technical devices and the social orders from which they emerge.

As there is an ecology to exercises of power, so is there an ecology to the manner in which we, as humans, build ICT infrastructures and their accompanying devices into our landscapes. This research has given me the opportunity to learn how ICTs and exercises of tribal sovereignty shape and are constrained by the policies, histories, and landscapes demarcating Indian Country. By framing contemporary build-outs of tribal broadband infrastructures against the history of US colonization, we, as information scientists, can appreciate why the United Nations General Assembly determined in 2012 that affordable Internet access is a human right, critical to democratic citizen government participation.[4] Affordable and robust broadband infrastructures are critical to self-governance and the exercise of sovereignty in Indian Country. Advocates for tribal media and broadband are working now to make sure that US federal authorities adjust policies and practices so that tribal leaders can weave this critical infrastructure into their homelands in accordance with community needs.

It is particularly important that tribal leaders retain command of the design, implementation, and maintenance of broadband networks. Commanding the build-out process, hardware decisions, network upgrades, content management, policies, and the work of personnel gives tribes the ability to evaluate the timing of upgrades, new enterprise, and agenda setting at a national level. Tribes do not have to own their own infrastructure or sponsor their own telecommunications companies and Internet service providers, but they do have to understand the uses and regulation of the infrastructure and services, including awareness of community information flows. They do have to build supportive and entrepreneurial telecom and Internet teams that will provide 24/7 Internet service to their communities, and they also need the wisdom to build in safeguards against the kinds of exploitative and predatory practices that beset their particular tribal communities. In that sense, making decisions about broadband policies, infrastructures, uses of ICTs, and information-sharing practices within the ecology of the reservation imbues broadband infrastructures with meaning. In Indian Country, the soil is not without meaning; even dust, heat, and rain clouds bear the significance of cultural sovereignty. The organizers of Idle No More conscripted smartphones, digital cameras, and laptops into the labor of Indigenous political mobilization. While there is presently a dearth of data on both the availability of these devices and broadband coverage rates in Indian Country, it is apparent that the devices and connectivity are present, impactful, and useful for governance goals.

Above all, this research shows that weaving broadband infrastructures into Indian Country is not just for general purposes of education, enterprise, or entertainment but is also about educating younger generations in Native, Indigenous, and tribal ways of knowing, supporting tribal enterprise, and encouraging the creativity of Native peoples in spite of colonization. Native youth have a consciousness—an awareness and sensation of exigency—and this Indigenous consciousness is the basis of future exercises of cultural sovereignty. As inherently sovereign peoples, this is the reason for our right to know.[5] As Native educators, we have a responsibility to educate future youth in the ways of our survival and continuation. As social scientists, we can ask, What new kinds of theories, design approaches, systems and devices do we need to support the long-term goals of Native and Indigenous peoples? What can we learn from their experiences of colonization? What does this teach us about ourselves, working to create deeper understandings of the impact of digital technologies and information in our everyday lives?

NOTES

INTRODUCTION

1 Febna Caven, "Being Idle No More: The Women behind the Movement," *Cultural Survival Quarterly* 37, no. 1 (2013): 6–7; Fyre Jean Graveline, "Idle No More: Enough Is Enough!" *Canadian Social Work Review* 29, no. 2 (2012): 293–300; Sonja John, "Idle No More: Indigenous Activism and Feminism," *Theory in Action* 8, no. 4 (2015): 38–54.

2 Luis Hernandez Navarro, "Zapatistas Can Still Change the Rules of Mexico's Politics," *Guardian*, December 31, 2012, www.theguardian.com/commentisfree/2012/dec/31/zapatistas-mexico-politics-protest (accessed August 3, 2016); El Kilombo, "We Don't Know You? EZLN Communiqué, December 29, 2012," *El Kilombo Intergaláctico*, January 2, 2013, www.elkilombo.org/we-dont-know-you-ccri-ezln-communique-december-29-2012/#_ftnref1 (accessed August 3, 2016); Tim Russo, "Zapatista March: The Deafening Silence of Resurgence," *Upside Down World: Covering Activism and Politics in Latin America*, December 22, 2012, http://upsidedownworld.org/main/mexico-archives-79/4041-zapatista-march-the-defeaning-silence-of-resurgence (accessed August 3, 2016).

3 "Yaquis mantienen bloqueada una carretera de Sonora," *Diario de Yucatán*, December 27, 2012, http://yucatan.com.mx/mexico/yaquis-mantienen-bloqueada-una-carretera-de-sonora (accessed August 3, 2016); Brenda Norrell, "Yaqui Vicam Pueblo International Gathering for the Defense of Water," *Censored News*, November 21, 2012, http://bsnorrell.blogspot.com/2012/11/yaqui-vicam-pueblo-international.html (accessed August 3, 2016).

4 Many thanks to Cree scholar and community health practitioner Shawn Wilson, author of *Research Is Ceremony*, for his presentation on relationality at the 2010 Native Organization of Indigenous Scholars annual symposium at the University of Washington, Seattle. Many Native and Indigenous peoples live through principles of interconnectedness. These principles help form the basis of many Native and Indigenous ways of knowing and philosophies. Earlier Indigenous thinkers, including Gregory Cajete and Megan Bang and Douglas Medin, have shown how principles of interconnectedness inform Indigenous approaches to environmental knowledge (Cajete, *Native Science: Natural Law of Interdependence* [Santa Fe, N.M.: Clear Light Publishers, 2000]; Bang and Medin, *Who's Asking?: Native Science, Western Science, and Science Education* [Cambridge, Mass.: MIT Press, 2014]). In some ways, the entrepreneurial realization of the value of Indigenous peoples' specific, rigorous, place-based, and relational understanding of dynamic living biomes has spurned the neoliberal project of classifying and evaluating Traditional Ecological Knowledge as both an economic and a scientific enterprise. However, in the spring of 2010, when Professor Wilson described relationality to us as a way of accepting the responsibility a whole person has to people, places, objects, ideas, and words in everyday life, I began to see how concepts of interconnectedness and relationships are also the basis for understanding the way Indigenous communiqués about resilience and possibility become internalized and enacted within communities. Bruno Latour and John Law provided the language for understanding how the values of a program ripple

through the process of its design, manifestation, and enactment (Latour, "Technology Is Society Made Durable," in *A Sociology of Monsters: Essays on Power, Technology, and Domination*, ed. John Law [London: Routledge, 1991], 103–131; Law, "Strategies of Power: Power, Discretion, and Strategy" [London: Routledge, 1991], 165–191). Though actor network theory finds its origin in a very different genealogy of ideas than do Native and Indigenous concepts of relationality, reciprocity, and reverie, it is because of my own attentiveness to moments of inspiration and reverie—a Yaqui (Yoeme) attentiveness to the blossoming and unfolding that occur when worlds, *aniam*, come together—that I breathed an Indigenous reading into actor network theory. Readers may want to consult David Delgado Shorter and Muriel Painter for an anthropological reading of Yaqui (Yoeme) worldviews (Shorter, "Hunting for History in Potam Pueblo: A Yoeme (Yaqui) Indian Deer Dancing Epistemology," *Folklore* 118 [2007]: 283–307; Painter, *With Good Heart: Yaqui Ceremonies and Beliefs in Pascua Village* [Tucson: University of Arizona Press, 1986]), but at this moment in the published history of science and philosophy, I must insist on the somewhat ineffable nature of what happens during Yaqui spiritual practice. The ways have not yet been translated into English or Spanish, nor have they been written about with the same profundity of meaning as they are danced, lived, breathed, or prayed through in the communities, in particular during Cuaresma (Waema). Atheists and secular humanists may not appreciate the idea that the path to scientific discovery can involve a spiritual discipline at times, but perhaps we can agree that scientists experience moments of revelation.

5 Jennifer Tupper, "Social Media and the Idle No More Movement: Citizenship, Activism, and Dissent in Canada," *Journal of Social Science Education* 13, no. 4 (2014): 87–94.

6 Taiake Alfred, "Sovereignty," in *Sovereignty Matters: Locations of Contestation and Possibility in Indigenous Struggles for Self-Determination*, ed. Joanne Barker (Lincoln: University of Nebraska Press, 2005).

7 Manuel Castells, *The Power of Identity*, vol. 2, *The Information Age: Economy, Society, and Culture* (Malden, Mass.: Blackwell Press, 1997); Phil Howard, *The Internet and Islam: The Digital Origins of Dictatorship and Democracy* (New York: Oxford University Press, 2010); Majid Tehranian, *Global Communication and World Politics: Domination, Development and Discourse* (Boulder, Colo.: Lynne Reiner, 1999); James Tully, *Public Philosophy in a New Key* (New York: Cambridge University Press, 2008).

8 Vine Deloria, Jr., "The Right to Know" (paper presented at the US Department of the Interior, Office of Library and Information Services, Washington, D.C., 1978); Allison B. Krebs, "Native America's Twenty-First-Century Right to Know," *Archival Science* 12 (2012): 173–90.

9 For further reading on the concept of epistemic injustice, please see Miranda Fricker, *Epistemic Injustice: Power and the Ethics of Knowing* (Cambridge: Oxford University Press, 2007); José Medina, *The Epistemology of Resistance: Gender and Racial Oppression, Epistemic Injustice, and Resistant Imaginations* (Cambridge: Oxford University Press, 2012); and Shannon Sullivan and Nancy Tuana, eds., *Race and Epistemologies of Ignorance* (New York: SUNY Press, 2007).

CHAPTER 1: NETWORK THINKING

1 Frederick Jackson Turner, "The Significance of the Frontier in American History" (paper presented at the World's Columbian Exposition, Chicago, 1893) (Washington, D.C.: American Historical Association, 1893).

2 Letter from Corporal Hervey Johnson to his sister Sybil, June 21, 1865, in William Unrau, *Tending the Talking Wire: A Buck Soldier's View of Indian Country, 1863–1866* (Salt Lake City: University of Utah Press, 1979), 261–64.

3 The eastern Sioux had already been in negotiations with neighboring European and American settlers for a long while before formal treaty agreements with the US federal government in the mid-1850s. Unfortunately, the government could not uphold the treaty obligations it had made, nor would it intervene in the unscrupulous behavior of traders who were withholding land payments and food supplies from the starving tribe. Violent skirmishes broke out between whites and Dakotas. White authorities viewed tribal people with no pity; they tried and executed thirty-eight Dakota men at one time in Mankato, Missouri. The year after, the US government pushed the eastern Sioux into Nebraska and South Dakota and eliminated the Minnesota Dakota reservation. President Lincoln referenced eight hundred whites killed during these battles, although there are no records of that number of whites or tribal people who died.

4 W. Bernard Carlson, *Tesla: Inventor of the Electrical Age* (Princeton, N.J.: Princeton University Press, 2013), 30.

5 L. G. Moses, *Wild West Shows and the Images of American Indians: 1883–1933* (Albuquerque: University of New Mexico Press, 1996); William Clements, *Imagining Geronimo: An Apache Icon in Popular Culture* (Albuquerque: University of New Mexico Press, 2013), 128–29.

6 Phil Deloria, *Indians in Unexpected Places* (Lawrence: University of Kansas Press, 2004.)

7 Silvia Federici, "Re-enchanting the World: Technology, the Body, and the Construction of the Commons," in *The Anomie of the Earth: Philosophy, Politics, and Autonomy in Europe and the Americas*, ed. Federico Luisetti, John Pickles, and Wilson Kaiser (Durham, N.C.: Duke University Press, 2015) , 202–14.

8 See the etymology of the word *information* in the *Oxford English Dictionary*, specifically, in *Paston Lett.* (1904) II. 38: "The said Erle..maye not..lette malicious and untrewe men to make informacions of his persone."

9 Jacques Ellul, *The Technological Society* (New York: Vintage Books, 1964).

10 Hans von Baeyer, *Information: The New Language of Science* (Cambridge, Mass.: Harvard University Press, 2004).

11 For more on the intersection of information, identity, and globalization, see Arun Agrawal, "Indigenous Knowledge and the Politics of Classification," *International Social Sciences Journal* 54, no. 173: 287–97; Manuel Castells, *The Rise of the Network Society*, vol. 1 of *The Information Age: Economy, Society and Culture* (Malden, Mass: Blackwell Press, 1996); Castells, *The Power of Identity*, vol. 2 of *The Information Age: Economy, Society and Culture* (Malden, Mass.: Blackwell Press, 1997); Castells, *The End of the Millennium*, vol. 3 of *The Information Age: Economy, Society and Culture* (Malden, Mass.: Blackwell Press, 1998); Marisa Elena Duarte, "Knowledge, Technology, and the Pragmatic Dimensions of Self-Determination," in *Restoring Indigenous Self-Determination*, ed. Marc Woons (Bristol, South West England: E-International Relations, 2014); Walter Mignolo, *The Darker Side of the Renaissance: Literacy, Territoriality, and Colonization* (Durham, N.C.: Duke University Press, 2003).

12 It is interesting to consider the impacts of the Net Neutrality debates within the frame of unserved and underserved reservation and rural communities. While corporate lobbyists seek to persuade political decision makers and consumers that greater access to bandwidth ought to come at a price, there are rural and reservation communities where people are already paying exorbitant rates for expensive, last-mile infrastructural build-

outs and very slow, and nearly unusable, Internet speeds. Meanwhile, programmers are developing apps, games, and systems that soak up more and more bandwidth. Economists are now calculating the relative value of spectrum. Meanwhile, free speech and public media advocates frame affordable high-speed Internet as a human right.

13 danah boyd, *It's Complicated: The Social Lives of Networked Teens* (New Haven, Conn.: Yale University Press, 2014).

14 Richard Coyne, *The Tuning of Place: Sociable Spaces and Pervasive Digital Media* (Cambridge, Mass.: MIT Press, 2010); Bonnie Nardi and Vicki O'Day, *Information Ecologies: Using Technology with Heart* (Cambridge, Mass.: MIT Press, 2000).

15 Manuel Castells, "A Network Theory of Power," *International Journal of Communication* 5 (2011): 773–87.

16 Miki Maaso, Felipe S. Molina, and Larry Evers, "The Elder's Truth: A Yaqui Sermon," *Journal of the Southwest* 35, no. 3 (1993): 225–317; Subcomandante Insurgente Marcos, "Other Intellectuals," in *The Speed of Dreams: Selected Writings 2001–2007*, ed. Canek Peña-Vargas and Greg Ruggiero (San Francisco: City Lights, 2006), 337–46.

17 Anna Munster, *An Aesthesia of Networks: Conjunctive Experience in Art and Technology* (Cambridge, Mass.: MIT Press, 2013), 5.

18 Maria Yellow Horse Brave Heart, "Models for Healing, Indigenous Survivors of Historical Trauma, Theory and Research Implications for the Seattle Community" (lecture presented at the Indigenous Wellness Research Institute, University of Washington, Seattle, October 2009).

19 Richard Slotkin, *Regeneration through Violence: The Mythology of the American Frontier, 1600–1860* (Oklahoma City: University of Oklahoma Press, 2000).

20 For a social scientific account of patterns in the methods of colonization, I have relied a great deal on work around the coloniality of power, formulated in Anibal Quijano, "Coloniality and Modernity/Rationality," *Cultural Studies* 21, no. 2 (2007): 168–78.

21 Ron Niezen, *The Origins of Indigenism: Human Rights and the Politics of Identity* (Berkeley: University of California Press, 2003), 12.

22 Jacques Ellul, *The Technological Society* (New York: Vintage Books, 1964).

23 Ibid.; Ellul, *The Technological System* (New York: Continuum, 1980).

24 Anibal Quijano, "Coloniality and Modernity/Rationality," *Cultural Studies* 21, no. 2 (2007): 168–78; Quijano, "Colonialidad del poder y clasificacion social," *Journal of World-Systems Research* 11, no. 2 (2000): 342–87; Quijano, "Coloniality of Power, Eurocentrism, and Latin America," *Nepantla: Views from the South* 1, no. 3 (2000): 549–54; Walter Mignolo, *The Darker Side of the Renaissance: Literacy, Territoriality, and Colonization* (Durham, N.C.: Duke University Press, 2003); Mignolo, *Local Histories / Global Designs: Coloniality, Subaltern Knowledges, and Border Thinking* (Princeton, N.J.: Princeton University Press, 2012).

25 Néstor García Canclini, *Imagined Globalization* (Durham, N.C.: Duke University Press, 2014); Federico Luisetti, John Pickles, and Wilson Kaiser, eds., *The Anomie of the Earth: Philosophy, Politics, and Autonomy in Europe and the Americas* (Durham, N.C.: Duke University Press, 2015); Walter Mignolo, *Local Histories / Global Designs: Coloniality, Subaltern Knowledges, and Border Thinking* (Princeton, N.J.: Princeton University Press, 2012); Sandro Mezzadra and Brett Nielson, *Border as Method, or the Multiplication of Labor* (Durham, N.C.: Duke University Press, 2013); Mabel Moraña, Enrique Dussel, and Carlos A. Jáuregi, eds., *Coloniality at Large: Latin America and the Postcolonial Debate* (Durham, N.C.: Duke University Press, 2008).

26 Christopher Bowman, "Indian Trust Fund: Resolution and Proposed Reformation to the

Mismanagement Problems Associated with the Individual Indian Money Accounts in Light of *Cobell v. Norton*," *Catholic University Law Review* 53, Cath. U. L. Rev. 543 (2004): 543–1195; Leslie Oakes and Joni Young, "Reconciling Conflict: The Role of Accounting in the American Indian Trust Fund Debacle," *Critical Perspectives on Accounting* 21, no. 1 (2010): 63–75.

27 Joseph Bock, *The Technology of Nonviolence: Social Media and Violence Prevention* (Cambridge, Mass.: MIT Press, 2012); Christian Fuchs, *Social Media: A Critical Introduction* (London: Sage, 2014); Phil Howard, *The Internet and Islam: The Digital Origins of Dictatorship and Democracy* (New York: Oxford University Press, 2010); Phil Howard and Muzzamil Mohammed Hussain, *Democracy's Fourth Wave? Digital Media and the Arab Spring* (London: Oxford, 2013); Karine Nahon and Jeff Helmsley, *Going Viral* (Cambridge: Polity, 2013).

28 Sonja John, "Idle No More: Indigenous Activism and Feminism," *Theory in Action* 8, no. 4 (2015): 38–54.

29 Susan Leigh Star and Anselm Strauss, "Layers of Silence, Arenas of Voice: The Ecology of Visible and Invisible Work," *Computer Supported Cooperative Work* 8, nos. 1/2 (1999): 9–30.

30 Audra Simpson, "Mapping Sovereignty: Indigenous Borderlands" (paper presented at "B/ordering Violence: Boundaries, Gender, Indigeneity in the Americas: 2012–13 John E. Sawyer Seminar in Comparative Cultures," Simpson Center for the Humanities, University of Washington, Seattle, April 11, 2013); Simpson, *Mohawk Interruptus: Political Life across the Borders of Settler States* (Durham, N.C.: Duke University Press, 2014).

CHAPTER 2: REFRAMING ICTS IN INDIAN COUNTRY

1 Makere Stewart-Harawira, *The New Imperial Order: Indigenous Responses to Globalization* (London: Zed Books, 2005).

2 Manuel de Landa, *A Thousand Years of Nonlinear History* (Cambridge, Mass.: MIT Press, 2000).

3 Vine Deloria, Jr., "Traditional Technology," in *Spirit and Reason: A Vine Deloria, Jr., Reader* (Boulder, Colo.: Fulcrum Press, 1999), 129–36.

4 Harry Cleaver, "The Zapatista Effect: The Internet and the Rise of an Alternative Political Fabric," *Journal of International Affairs* 51, no. 2 (1998): 620–40; Maria Garrido, "The Zapatista Indigenous Women: The Movement within the Movement" (lecture presented at the Department of Gender and Women's Studies, University of Washington, Seattle, 2011); Maria Elena Martinez-Torres, "Civil Society, the Internet, and the Zapatistas," *Peace Review* 13, no. 3 (2001): 347–55; Sheryl Shirley, "Zapatista Organizing in Cyberspace: Winning Hearts and Minds?" (paper presented at the Conference of the Latin American Studies Association, Washington, D.C., 2001).

5 Channette Romero, "Envisioning a 'Network of Tribal Coalitions': Leslie Marmon Silko's *Almanac of the Dead*," *American Indian Quarterly* 26, no. 4 (2002): 623–40.

6 James Casey, Randy Ross, and Marcia Warren, *Native Networking: Telecommunications and Information Technology in Indian Country* (Washington, D.C.: Benton Foundation, 1999); Traci Morris and Sascha Meinrath, *New Media, Technology, and Internet Use in Indian Country: Quantitative and Qualitative Analyses* (Phoenix, Ariz.: Native Public Media; Washington, D.C.: New America Foundation, 2009); Office of Technology Assessment, US Congress, *Telecommunications Technology and Native Americans: Opportunities and Challenges*, OTA-ITC-621 (Washington, D.C.: US Government Printing Office,

1995); Linda Ann Riley, Bahram Nassarsharif, and John Mullen, *Assessment of Technology Infrastructure in Native Communities*, Economic Development Administration (Las Cruces: New Mexico State University, 1999).

7 Jessica Dorr and Richard Akeroyd, "New Mexico Tribal Libraries: Bridging the Digital Divide," *Computers in Libraries* 21, no. 8 (2001): 8; Laurel Evelyn Dyson, Max Hendriks, and Stephen Grant, *Information Technology and Indigenous People* (Hershey, Penn.: Information Science Publishing, 2007); Andrew Gordon, Margaret Gordon, and Jessica Dorr, "Native American Technology Access: The Gates Foundation in Four Corners," *Electronic Library* 21, no. 5 (2001): 428–34; Richard McMahon, "The Institutional Development of Indigenous Broadband Infrastructure in Canada and the United States: Two Paths to Digital 'Self-Determination,'" *Canadian Journal of Communication* 36 (2011): 115–40; Traci Morris and Sascha Meinrath, *New Media, Technology, and Internet Use in Indian Country: Quantitative and Qualitative Analyses* (Phoenix, Ariz.: Native Public Media; Washington, D.C.: New America Foundation, 2009); Jayson Richardson and Seon McLeod, "Technology Leadership in Native American Schools," *Journal of Research in Rural Education* 26, no. 7 (2011): 1–14; James Stevens, "E-Socials: Cultural Collaboration in the Age of the Electronic Inter-Tribal," in *Sovereign Bones: New Native American Writing*, ed. Eric Gansworth (New York: Nation Books, 2007).

8 Teresa Bissell, "The Digital Divide Dilemma: Preserving Native American Culture While Increasing Access to Information Technology on Reservations," *Journal of Law, Technology, and Policy*, 1 (2004): 129–50; Kathleen Buddle, "Aboriginal Cultural Capital Creation and Radio Production in Urban Ontario," *Canadian Journal of Communication* 30, no. 1 (2005): 7–39; Jeremy Busacca, "Seeking Self-Determination: Framing, the American Indian Movement, and American Indian Media" (PhD diss., Claremont Graduate University, 2007); Mac Chapin, Zachary Lamb, and Bill Threlkeld, "Mapping Indigenous Lands," *Annual Review of Anthropology* 34 (2005): 619–38; Ross Frank, "The Tribal Digital Village: Technology, Sovereignty, and Collaboration in Indian Southern California" (unpublished manuscript, University of California, San Diego, 2004), http://pages.ucsd.edu/~rfrank/class_web/ETHN200C/TDVchapters.pdf; Jason Heppler, "Framing Red Power: The American Indian Movement, the Trail of Broken Treaties, and the Politics of Media" (PhD diss., University of Nebraska, Lincoln, 2009); Kyra Landzelius, *Native on the Net: Indigenous and Diasporic Peoples in the Virtual Age* (London: Routledge, 2006); Jerry Mander, *The Failure of Technology and the Survival of the Indian Nations* (San Francisco: Sierra Club Books, 1991); Ramesh Srinivasan, "Tribal Peace: Preserving the Cultural Heritage of Dispersed Native American Communities" (paper presented at the International Conference on Cultural Heritage and Informatics, Berlin, September 2004); Pamela Wilson and Michelle Stewart, *Global Indigenous Media: Cultures, Poetics, and Politics* (Durham, N.C.: Duke University Press, 2008).

9 Jim Ereaux, "The Impact of Technology on Salish Kootenai College," *Wicazo Sa Review* 13, no. 2 (1998): 117–35; Jerry Mander, *The Failure of Technology and the Survival of the Indian Nations* (San Francisco: Sierra Club Books, 1991); Jean- François Savard, "A Theoretical Debate on the Social and Political Implications of the Internet for the Inuit of Nunavut," *Wicazo Sa Review* 13, no. 2 (1998): 83–97.

10 Arthur Kroker, *The Will to Technology and the Culture of Nihilism: Heidegger, Nietzsche, Marx* (Toronto: University of Toronto Press, 2004).

11 Gregory Cajete, *Native Science: Natural Law of Interdependence* (Santa Fe, N.M.: Clear Light Publishers, 2000); Douglas Medin and Megan Bang, *Who's Asking?: Native Science, Western Science, and Science Education* (Cambridge, Mass.: MIT Press, 2014).

12 Miranda Fricker, *Epistemic Injustice: Power and the Ethics of Knowing* (New York: Oxford University Press, 2009); José Medina, *The Epistemology of Resistance: Oppression, Epistemic Injustice, and Resistant Imaginations* (Cambridge: Oxford University Press, 2012); Shannon Sullivan and Nancy Tuana, eds., *Race and Epistemologies of Ignorance* (Buffalo: SUNY Press, 2012).

13 Linda Tuhiwai Smith, *Decolonizing Methodologies: Research and Indigenous Peoples* (London: Zed Books, 1999).

14 Noel Dyck, *What Is the Indian Problem? Tutelage and Resistance in Canadian Indian Administration* (St. Johns, Newfoundland, Canada: Institute of Social and Economic Research, 1991); Robert Hays, *Editorializing "The Indian Problem": The New York Times on Native Americans, 1860–1900* (Carbondale: Southern Illinois University Press, 2007).

15 Marisa Elena Duarte, "Connected Activism: Indigenous Uses of Social Media for Shaping Political Change" (paper presented at "By The People: Participatory Democracy, Civic Engagement, and Citizenship Education," Tempe, Ariz., December 3, 2015).

16 Manuel Castells, *The Rise of the Network Society*, vol. 1 of *The Information Age: Economy, Society, and Culture* (Malden, Mass.: Blackwell, 1996).

17 Manuel Castells, *The Power of Identity*, vol. 2 of *The Information Age: Economy, Society, and Culture* (Malden, Mass.: Blackwell, 1997).

18 Gino Orticio, *Indigenous/Digital Heterogeneities: An Actor-Network-Theory Approach* (doctoral thesis, Queensland University of Technology, 2013); Eva Silven, "Contested Sami Heritage: Drums and Sieidis on the Move" (paper presented at "EuNaMus 2012, Identity Politics, the Uses of the Past, and the European Citizen," Brussels, January 26–27, 2012); Vanessa Watts, "Indigenous Place-Thought and Agency amongst Humans and Non-humans (First Woman and Sky Woman Go on a European World Tour!)," *Decolonization: Indigeneity, Education, and Society* 2, no. 1 (2013): 20–34.

19 Wiebe Bijker and John Law, *Shaping Technology/Building Society: Studies in Sociotechnical Change* (Cambridge, Mass.: MIT Press, 1992); Bruno Latour, *Reassembling the Social: An Introduction to Actor-Network-Theory* (Oxford: Oxford University Press, 2007); John Law, *A Sociology of Monsters: Essays on Power, Technology, and Domination* (New York: Routledge, 1991).

20 Chet Bowers, Miguel Vasquez, and Mary Roaf, "Native Peoples and the Challenge of Computers: Reservation Schools, Individualism, and Consumerism," *American Indian Quarterly* 24, no. 2 (2000): 182–99; Craig Howe, "Cyberspace Is No Place for Tribalism," *Wicazo Sa Review* 13, no. 2 (1998): 19–28; Loriene Roy, "Four Directions: An Indigenous Educational Model," *Wicazo Sa Review* 13, no. 2 (1998): 59–69; Michael Two Horses, "Gathering around the Electronic Fire: Persistence and Resistance in Electronic Formats," *Wicazo Sa Review* 13, no. 2 (1998): 29–43.

CHAPTER 3: THE OVERLAP BETWEEN TECHNOLOGY AND SOVEREIGNTY

1 Vine Deloria, Jr., "Traditional Technology," in *Spirit and Reason: A Vine Deloria Jr. Reader* (Boulder, Colo.: Fulcrum Press, 1999), 129–36.

2 To get a sense of the conflicting and colonial attitudes and worldviews shaping outcomes of the Havasupai case, see Nanibaa Garrison, "Genomic Justice for Native Americans: Impact of the Havasupai Case on Genetic Research," *Science, Technology & Human Values* 38, no. 2 (2013): 201–23.

3 Ibid.; Amy Harmon, "Indian Tribe Wins Fight to Limit Research of Its DNA," *New York Times*, April 21, 2010; Havasupai Tribe of the Havasupai Reservation v. Arizona Board

of Regents and Therese Ann Markow, Arizona Court of Appeals nos. 1 CA-CV 07-0454, 1 CA-CV 07-0801 (2008).

4 Mark Shaffer, "Havasupai Blood Samples Misused," *Indian Country Today*, March 9, 2004; Michael Kiefer, "Havasupai Ends Regents Lawsuit with Burial," *Arizona Republic*, April 22, 2010.

5 Amanda Cobb, "Understanding Tribal Sovereignty: Definitions, Conceptualizations, and Interpretations," *American Studies* 46, nos. 3/4 (2005): 115–32; Vine Deloria, Jr., and Clifford Lytle, *American Indians, American Justice* (Austin: University of Texas Press, 1983); Taiake Alfred, "Sovereignty," in *Sovereignty Matters: Locations of Contestation and Possibility in Indigenous Struggles for Self-Determination*, ed. Joanne Barker (Lincoln: University of Nebraska Press, 2005), 33–50.

6 Marisa Elena Duarte, Miranda Belarde-Lewis, and Allison B. Krebs, "Native Systems of Knowledge: Indigenous Methodologies in Information Science" (lecture presented at the iConference, Urbana, Ill., 2010); Tom Holm, Diane Pearson, and Ben Chavis, "Peoplehood: A Model for the Extension of Sovereignty in American Indian Studies," *Wicazo Sa Review* 18, no. 1 (2003): 7–24; Allison B. Krebs, "Native America's Twenty-First-Century Right to Know," *Archival Science* 12 (2012): 173–90; Krebs, "Indigenous Information Ecology: Vanishing Indians Throwing Off Our Invisibility Cloaks Rushing into the 21st Century" (lecture presented at the School of Information Resources and Library Science, University of Arizona, Tucson, March 10, 2008).

7 LaDonna Harris et al., "Returning to Harmony through the Wisdom of the People: Applying Traditional Principles to Develop Appropriate and Effective Indian Tribal Governance; Returning Indian Nations to Culturally Appropriate Forms of Decision Making," in *Re-creating the Circle: The Renewal of Indian Self-Determination*, ed. LaDonna Harris, Stephen Sachs, and Barbara Morris (Albuquerque: University of New Mexico Press, 2011), 201–50; Sarah Hicks, "Intergovernmental Relationships: Expressions of Tribal Sovereignty," in *Rebuilding Native Nations: Strategies for Tribal Governance*, ed. Miriam Jorgensen (Tucson: University of Arizona Press, 2007), 246–72.

8 The arguments precluding tribal access to spectrum are complicated and recurrent. At present, network engineers and scientists in Indian Country are still seeking support—both from national science funding sources and through federal subsidy—to design and deploy digital signals across ultrawideband radio spectrum over reservation land. For a sense of how complicated this issue is with regard to the sovereign rights of tribes—not only for radio spectrum but also for broadband spectrum—see John C. Miller and Christopher P. Guzelian, "The Spectrum Revolution: Deploying Ultrawideband Technology on Native American Lands," *11 Commlaw Conspectus* 277 (2003): 277–305. It is no small contribution that Hector Youtsey and the Pascua Yaqui Tribe radio staff made in advocating for access to spectrum for Native American reservations.

9 For a more detailed account, see Evelyn Hu-Dehart, *Missionaries, Miners, and Indians: History of Spanish Contact with the Yaqui Indians of Northwestern New Spain, 1533–1830* (Tucson: University of Arizona Press, 1981); Hu-Dehart, *Yaqui Resistance and Survival: Struggle for Land and Autonomy, 1821–1910* (Madison: University of Wisconsin Press, 1984); Edward Spicer, *Cycles of Conquest: The Impact of Spain, Mexico, and the United States on Indians of the Southwest, 1533–1960* (Tucson: University of Arizona Press, 1967); Hector Cuauhtemoc Hernandez Silva, *Insurgencia y autonomía: Historia de los pueblos Yaquis, 1821–1910* (Ciudad de México: Instituto Nacional Indigenista, 1996).

10 United States Customs and Border Patrol, *US Customs and Border Protection Performance*

and Accountability Report, Fiscal Year 2008 (Washington, D.C.: Office of Finance, 2008). United States Government Accountability Office, *Alien Smuggling: DHS Needs to Better Leverage Investigative Resources and Measure Program Performance along the Southwest Border: Report to Congressional Requesters* (GAO-10-2328, 2010).

11 Jason De León and Mike Wilson, "Rights, Sovereignty, and Lives on the Line: Immigration Debates across Arizona and Tohono O'odham Borderlands" (lecture presented at the Center for Global Studies, University of Washington, Seattle, February 2010).

12 Ibid.

13 Taiake Alfred, "Sovereignty," in *Sovereignty Matters*, ed. Joanne Barker (Lincoln: University of Nebraska Press, 2005), 33–50.

14 Subcommittee on Management, Integration and Oversight of the Committee on Homeland Security, House of Representatives, *The Secure Border Initiative: Ensuring Effective Implementation and Financial Accountability of SBInet: Hearing before the Subcommittee on Management, Integration, and Oversight of the Committee on Homeland Security*, 109th Cong., 2nd sess., November 15, 2006; United States Government Accountability Office, *Secure Border Initiative: DHS Needs to Reconsider Its Proposed Investment in Key Technology Program* (GAO-10-340, 2010).

15 Oliphant v. Suquamish 435 U.S. 191 (1978). At the time of this writing, Congress has passed the Violence against Women Act, and my own tribe, the Pascua Yaqui Tribe, and a geographic neighbor to the Tohono O'odham Tribe, is one of the tribes piloting the act. In time we will understand how the Violence against Women Act shapes the ability of tribal law enforcement officials to apprehend, and of tribal judges to try, non-tribal criminal suspects on tribal lands.

16 Charles Bowden, *Blue Desert* (Tucson: University of Arizona Press, 1986); Bowden, *Murder City: Ciudad Juarez and the Global Economy's New Killing Fields* (New York: Nation Books, 2010); Richard Drinnon, *Facing West: The Metaphysics of Indian-Hating* (Oklahoma City: University of Oklahoma Press, 1997); Nicole Guidotti-Hernandez, *Unspeakable Violence: Remapping U.S. and Mexican National Imaginaries* (Durham, N.C.: Duke University Press, 2011); Karl Jacoby, *Shadows at Dawn: An Apache Massacre and the Violence of History* (New York: Penguin Books, 2009); Richard Slotkin, *Regeneration through Violence: The Mythology of the American Frontier, 1600–1860* (Oklahoma City: University of Oklahoma Press, 2000).

17 Gavin Clarkson, Jacob Trond, and Archer Batcheller, "Information Asymmetries and Information Sharing," *Government Information Quarterly* 24 (2007): 827–39; Debbie Conner, "'To the Most Exacting Fiduciary Standards': Notes on Information Technologies and Federal Administration of Indian Trust Lands and Resources," *Wicazo Sa Review* 13, no. 2 (1998): 99–116.

18 In my own tribe, in Pascua during fiestas, the *cabos*—the guards who maintain social order for the ceremonial environment—remind people to leave the area if they are using mobile phones. If individuals are caught taking pictures or recording prayer events, the *cabos* confiscate the phones and delete the photos. The tribe posts signs with clear wording: no photos, sketching, or note taking is allowed. The debate over uses of recording technologies in ceremonial places pivots around a few important points: respect for being in the present in prayer; respect for the privacy of individuals fulfilling their *mandas* and *ofisios*, or sacred obligations; prevention of exploitative behavior by amateur and/or unethical ethnographers seeking to "take," document, repackage, and profit from Yoeme philosophical and spiritual practices without the active participation and permission of elders; concerns about the effect of mediation on powerful spiritual prac-

tices; and concerns that younger generations will forsake ceremonial rhythms and order for the artificial time and interfaces of textual digital culture.

19 Miranda Belarde-Lewis, "Sharing the Private in Public: Indigenous Cultural Property in Online Media" (paper presented at the iConference, Seattle, February 2011).

20 James Casey, Randy Ross, and Marcia Warren, *Native Networking: Telecommunications and Information Technology in Indian Country* (Washington, D.C.: Benton Foundation, 1999); Jessica Dorr and Richard Akeroyd, "New Mexico Tribal Libraries: Bridging the Digital Divide," *Computers in Libraries* 21, no. 8 (2001): 8; Evelyn Dyson, Max Hendriks, and Stephen Grant, *Information Technology and Indigenous People* (Hershey, Pa.: Information Science Publishing, 2007); Andrew Gordon, Margaret Gordon, and Jessica Dorr, "Native American Technology Access: The Gates Foundation in Four Corners," *Electronic Library* 21, no. 5 (2001): 428–34; Jerry Mander, *The Failure of Technology and the Survival of the Indian Nations* (San Francisco: Sierra Club Books, 1991); Office of Technology Assessment, US Congress, *Telecommunications Technology and Native Americans: Opportunities and Challenges*, OTA-ITC-621 (Washington, D.C.: US Government Printing Office, 1995); Linda Ann Riley, Bahram Nassarsharif, and John Mullen, *Assessment of Technology Infrastructure in Native Communities*, Economic Development Administration (Las Cruces: New Mexico State University, 1999).

21 Vine Deloria, Jr., "Traditional Technology," in *Spirit and Reason: A Vine Deloria, Jr., Reader* (Boulder, Colo.: Fulcrum Press, 1999), 129–36; Martin Heidegger, *The Question Concerning Technology and Other Essays* (New York: Harper Colophon, 1977).

22 It is most interesting to read Heidegger's "Question Concerning Technology" from an Indigenous perspective. Heidegger wrote the lecture in the late 1940s, during the years that he was banned from teaching due to his affiliation with the Nazi (National Socialist) party during his early career as a rector at the University of Freiburg. In the postwar period, he faced criticism for the essays and lectures he prepared during his banishment because of the associations he drew between industrialization and the Nazi administration's inhumane treatment of people. Heidegger acknowledges the inner destructiveness that certainly must ensue should modern European man invest in technological advance as his most natural mode of transcendence. Continually realizing himself through the design of a social order based around the actions of technique, categorization, reduction, synthesis, and mechanization, the modern man begins to render the world around him as a process of function. Heidegger worries about the loss of a true aesthetics, a true art that is founded in appreciation of nonmechanical, nontechnological domains of knowledge: the natural, the ephemeral, the ideal. He refers to the need for "reserves," places where the technological does not rule through totality, through the eclipse of reverie by technique. He claims that inspiration and art can emerge only through the "reserve." At the time that Heidegger was thinking in these terms, the roots of the American dream—Manifest Destiny—had already emerged out of the European claim to the New World, feeding the subsequent Euro-American industrialization toward eventual global market domination. Native and Indigenous peoples were being placed on reserves and reservations, and their participation as epistemic partners in all of the institutions of industrial advance—business, colleges and universities, banking and finance industries, railroads, telecommunications, light and power, and the various accompanying associations—was being curbed. Indians were treated as antitechnological and antimodern, as more fauna than human and, often, as vermin. At present, I regularly attend meetings of technologists and scientists who rely on digital technology to accomplish their research designs and data collection, and many ask how we can merge

Indigenous understandings of ecology and the environment in order to identify technologies for sustainable living. I think of Heidegger's "Question Concerning Technology" and the society in which he lived. The politics of Indigeneity—how Indigeneity came to be in many ways outside of and in assertive response to the deleterious technology and power of states—cannot be separated from Native and tribal ways of knowing about the rhythms of the natural world. From the reserve, the reservation, we have to interrogate the Euro-American belief in national progress through techno-scientific advance by asking, Who is this technology for? To what end? And how?

CHAPTER 4: SOCIOTECHNICAL LANDSCAPES

1 Paul Dourish and Genevieve Bell, *Divining a Digital Future: Mess and Mythology in Ubiquitous Computing* (Cambridge, Mass.: MIT Press, 2011).
2 Geoffrey Bowker and Susan Leigh Star, *Sorting Things Out: Classification and Its Consequences* (Cambridge: MIT Press, 1999), 10.
3 Mac Chapin, Zachary Lamb, and Bill Threlkeld, "About Us," Cheyenne River Sioux Tribe Telephone Authority, http://crsta.com/about/about-us.php (accessed January 10, 2013).
4 John Law, "Power, Discretion, and Strategy," in *A Sociology of Monsters: Essays on Power, Technology, and Domination*, ed. John Law (London: Routledge, 1991), 165–91; Bruno Latour, "Technology Is Society Made Durable," in *A Sociology of Monsters: Essays on Power, Technology, and Domination*, ed. John Law (London: Routledge, 1991), 103–31.
5 Richard Coyne, *The Tuning of Place: Sociable Spaces and Pervasive Digital Media* (Cambridge, Mass.: MIT Press, 2010).

CHAPTER 5: INTERNET FOR SELF-DETERMINATION

1 William J. Bauer, Jr., "When the Owens Valley Went Dry: Politics, Water, and the Paiute Oral Tradition in the 1930s" (paper presented to the Native American and Indigenous Studies Association, Sacramento, Calif., 2011).
2 National Telecommunications and Information Administration and Federal Communications Commission, "National Broadband Map (2013)," www.broadbandmap.gov (accessed June 5, 2013). The challenges associated with using global information system mapping and working with the federal government to map tribal lands are laden with paternalistic attitudes and practices. See Mark Palmer, "Cartographic Encounters at the Bureau of Indian Affairs Geographic Information System Center of Calculation," *American Indian Culture and Research Journal* 36, no. 2 (2012): 75–102.
3 Silvia Federici, "Re-enchanting the World: Technology, the Body, and the Construction of the Commons," in *The Anomie of the Earth: Philosophy, Politics, and Autonomy in Europe and the Americas*, ed. Federico Luisetti, John Pickles, and Wilson Kaiser (Durham, N.C.: Duke University Press, 2015), 202–14.
4 Margaret Noori, "Waasechibiiwaabikoonsing Nd'anami'aa, 'Praying through a Wired Window': Using Technology to Teach Anishinaabemowin," *Studies in American Indian Literatures* 23, no. 2 (2011): 18.
5 The world of tribal broadband advocacy is small. Traci Morris and Matt Rantanen had long recognized that this was an area of policy work that needed research, advocacy, and awareness efforts. As I was writing up the findings of my research, Morris was partnering with Miriam Jorgensen of the Native Nations Institute and with Susan Feller, a former National Science Foundation program officer and member of the board of the

Association of Tribal Archives, Libraries, and Museums, to research Internet uses in tribal libraries. In the study, they address the issue of e-rate funding for tribal schools and libraries. See Miriam Jorgensen, Traci Morris, and Susan Feller, *Digital Inclusion in Native Communities: The Role of Tribal Libraries* (Oklahoma City, Okla.: Association of Tribal Archives, Libraries, and Museums, 2014).

6 Federal Communications Commission, *Federal Communications Commission Office of Native Affairs and Policy 2012 Annual Report* (Washington, D.C.: FCC, 2012).

7 Indigenous Commission for Communications Technologies in the Americas, *The Plan: Indigenous Peoples Empowering Themselves through Technology* (Ottawa: ICCTA, 2009).

8 Red Spectrum Communications, Inc., "Coeur d'Alene Tribe's Fiber to the Home Project (2013)," http://redspectrum.com/fibermain.html (accessed April 12, 2013).

9 Ibid.

10 Cheyenne River Sioux Tribe Telephone Authority, Comments before the Federal Communications Commission, *In the Matter of Inquiry Regarding Current Carrier Systems, Including Broadband over Power Lines Systems (ET Docket No. 03-104), Amendment of Part 15 Regarding New Requirements and Measurement Guidelines for Access: Broadband over Power Line Systems, April 29, 2004 (ET Docket No. 04-37)*, available at https://ecfsapi.fcc.gov/file/6516182369.pdf (accessed November 5, 2016).

CHAPTER 6: NETWORK SOVEREIGNTIES

1 Richard Drinnon, *Facing West: The Metaphysics of Indian-Hating and Empire-Building* (Minneapolis: University of Minnesota Press, 1980); Thomas Richards, *The Imperial Archive: Knowledge and the Fantasy of Empire* (London: Verso, 1996).

2 Armen Merjian, "An Unbroken Chain of Injustice: The Dawes Act, Native American Trusts, and *Cobell v. Salazar*," *Gonzaga Law Review* 46 (2010): 609–60.

3 Vine Deloria, Jr., and Clifford Lytle, *American Indians, American Justice* (Austin: University of Texas Press, 1983); David Wilkins, *American Indian Sovereignty and the U.S. Supreme Court: The Masking of Justice* (Austin: University of Texas Press, 1997).

4 Federal Communications Commission, "FCC Chairman Genachowski Appoints Geoffrey Blackwell to Lead New Initiatives on Native Affairs," *FCC News, June 22, 2010*, http://hraunfoss.fcc.gov/edocs_public/attachmatch/DOC-298924A1.pdf (accessed June 13, 2013).

5 Federal Communications Commission, *Federal Communications Commission Office of Native Affairs and Policy 2012 Annual Report* (Washington, D.C.: FCC, 2012).

6 Vine Deloria, Jr., and Clifford Lytle, *American Indians, American Justice* (Austin: University of Texas Press, 1983).

7 Native American Broadband Association, "Native American Broadband Association, 2010," www.nativeamericanbroadband.org (accessed January 15, 2011).

8 Vine Deloria, Jr., "Traditional Technology," in *Spirit and Reason: A Vine Deloria, Jr., Reader* (Boulder, Colo.: Fulcrum Press, 1999); Deloria, Jr., "If You Think about It, You Will See That It Is True," in *Spirit and Reason: A Vine Deloria, Jr., Reader* (Boulder, Colo.: Fulcrum Press, 1999).

9 Federal Power Commission v. Tuscarora Indian Nation 362 U.S. 99 (1960).

10 Manuel Castells, *The Power of Identity*, vol. 2 of *The Information Age: Economy, Society and Culture* (Malden, Mass.: Blackwell Press, 1997).

11 Ron Niezen, *The Origins of Indigenism: Human Rights and the Politics of Identity* (Berkeley: University of California Press, 2003).

12 Michael Adas, *Machines as the Measure of Men: Science, Technology, and Ideologies of Western Dominance* (Ithaca, N.Y.: Cornell University Press, 1989).

13 Frederick Jackson Turner, "The Significance of the Frontier in American History" (paper presented at the World's Columbian Exposition, Chicago, 1893) (Washington, D.C.: American Historical Association, 1893).

14 Worcester v. Georgia 31 US (6 Pet.) 515 (1832); Lewis Mumford, *Technics and Civilization* (Chicago: The University of Chicago Press, 1934).

15 Pacific Railroad Acts, 12 § 489 (1862, 1863, 1864).

16 George Croffut, *Crofutt's New Overland Tourist and Pacific Coast Guide* (Omaha, Neb.: Overland, 1878).

17 Linda Tuhiwai Smith, *Decolonizing Methodologies: Research and Indigenous Peoples* (London: Zed Books, 1999).

18 Indian Self-Determination and Education Assistance Act of 1975 25 § 450; Telecommunications Act of 1996, Pub. LA. No. 104, 110 Stat. 56 (1996).

19 United States Congress, Native American Telecommunications Act of 1997, 105th Cong., 1997–98, H.R. 486.

20 John C. Miller and Christopher P. Guzelian, "The Spectrum Revolution: Deploying Ultra-wideband Technology on Native American Lands," *11 Commlaw Conspectus* 277 (2003): 289.

21 Frederick Jackson Turner, "The Significance of the Frontier in American History" (paper presented at the World's Columbian Exposition, Chicago, 1893) (Washington, D.C.: American Historical Association, 1893).

22 United Nations General Assembly, United Nations Declaration on the Rights of Indigenous Peoples: Resolution / Adopted by the General Assembly, October 2, 2007, A/RES/61/295.

23 Subcomandante Insurgente Marcos, "The First Other Winds," in *The Speed of Dreams: Selected Writings 2001–2007*, ed. Marco Canek Peña-Vargas and Greg Ruggiero (San Francisco: City Lights, 2007), 302–17.

24 For more on distinct digital imaginaries in the global south, in particular in Peru, see Anita Say Chan, *Networking Peripheries: Technological Futures and the Myth of Digital Universalism* (Cambridge, Mass.: MIT Press, 2014).

25 Amanda Cobb, "Understanding Tribal Sovereignty: Definitions, Conceptualizations, and Interpretations," *American Studies* 46, nos. 3–4 (2005): 115–32.

26 Indian Self-Determination and Education Assistance Act of 1975 25 § 450.

27 Digital Arizona Council, *Digital Arizona: Expanding Innovation through Connectivity: Arizona's Strategic Plan for Digital Capacity: Expanded and Reference Version* (Phoenix: Arizona Department of Administration, Arizona Strategic Enterprise Technology, 2012).

28 "Internet Group Ranks Montana among the 'Disconnected Dozen,'" *Associated Press State and Local Wire*, July 29, 1999.

29 Phillip Dampier, "Montana's Struggle for Broadband Pits Cable, Phone Companies, and Native American Communities against One Another," Stop the Cap! Promoting Better Broadband, Fighting Data Caps, Usage-Based Billing, and Other Internet Overcharging Schemes, February 10, 2010, http://stopthecap.com/2010/02/10/montanas-struggle-for-broadband-pits-cable-phone-companies-and-native-american-communities-against-one-another/ (accessed March 9, 2013).

30 "Mont. Stimulus-Funded Internet Expansion Assailed," *Associated Press State and Local Wire*, November 4, 2009.

31 Randy Evans and Jim Dunstan, *Communications Regulation and Taxation in Indian*

Country: Executive Summary: A Report Prepared in Conjunction with the Tribal Telecom 2013 Conference, Law Office of Randal T. Evans, PLLC, Phoenix, Ariz., and Mobius Legal Group, PLLC, Springfield, Va., 2013; Evans and Dunstan, *Tribal Broadband Guide: Telecommunications Regulation and Taxation in Indian Country* (Washington, D.C.: Native American Finance Officers Association, 2013).

32 William Jefferson Clinton, "Executive Order 13175: Consultation and Coordination with Indian Tribal Governments," *Federal Register* 65, no. 218 (2000): 67249–67251.

33 Federal Communications Commission, *Federal Communications Commission Office of Native Affairs and Policy 2012 Annual Report* (Washington, D.C.: FCC, 2012).

34 Dawes Act, or General Allotment Act of 1887, 25 § 331.

35 Erik Cutright, "Tribal Telecom Enterprises and ISPs: Connecting ALL the Dots: Karuk Tribe" (lecture presented at TribalNet, San Diego, Calif., 2012).

36 J. D. Williams, *Statement by J.D. Williams, Cheyenne River Sioux Telephone Authority, before the Federal Communications Commission Regarding Overcoming Obstacles to Telephone Service to Indians on Reservations*, March 23, 1999.

37 Stewart Clegg, *Frameworks of Power* (London: Sage, 1989); Bruno Latour, "Technology Is Society Made Durable," in *A Sociology of Monsters: Essays on Power, Technology, and Domination*, ed. John Law (London: Routledge, 1991), 103–31; John Law, "Power, Discretion, and Strategy," in *A Sociology of Monsters: Essays on Power, Technology, and Domination*, ed. John Law (London: Routledge, 1991), 165–91.

38 Federal Communications Commission, *Federal Communications Commission Office of Native Affairs and Policy 2012 Annual Report* (Washington, D.C.: FCC, 2012).

CHAPTER 7: DECOLONIZING THE TECHNOLOGICAL

1 John Curran, "Governance, Communications Technologies, and the Transition to IPV6" (presentation at the Tribal Telecom and Technology Summit, Gila River, Ariz., February 2014).

2 Government Accountability Office, *Tribal Internet Access: Increased Federal Coordination and Performance Measurement Needed*, by Mark Goldstein, GAO 16-504T (Washington, D.C.: GAO, 2016.

3 I am certainly not the first Indigenous theorist attempting to bridge metaphors between Spider Woman, digital networks, and Native peoples' ways of creating new ways of being. In the first decade of the 2000s, Gabriel S. Estrada, while incorporating digital pedagogies into his courses at the University of Arizona, began conceptualizing NativeWeb, a digital portal created by and for Native educators, as an "electronic extension of Spider Woman's knowledge on the World Wide Web where all perception is interrelated and rapidly evolving beyond our comprehension." Gabriel Estrada, "Native Avatars, Online Hubs, and Urban Indian Literature," *Studies in American Indian Literatures* 23, no. 2 (2011): 49.

4 Eve Tuck and Wayne Yang, "Decolonization Is Not a Metaphor," *Decolonization: Indigeneity, Education, and Society* 1, no. 1 (2012): 1–40.

5 Franz Fanon, "Concerning Violence," *The Wretched of the Earth* (New York: Grove Press, 1963), 35–106.

6 Prasenjit Duara, *Decolonization: Perspectives from Now and Then* (New York: Routledge, 2004); Franz Fanon, *The Wretched of the Earth* (New York: Grove Press, 1963); Vijay Prashad, *The Darker Nations: A People's History of the Third World* (New York: The New

Press, 2008); Walter Mignolo, *Local Histories / Global Designs: Coloniality, Subaltern Knowledges, and Border Thinking* (Princeton, N.J.: Princeton University Press, 2012); Anibal Quijano, "Coloniality of Power, Eurocentrism, and Latin America," *Nepantla: Views from the South* 1, no. 3 (2000): 549–54; Eve Tuck and Wayne Yang, "Decolonization Is Not a Metaphor," *Decolonization: Indigeneity, Education, and Society* 1, no. 1 (2012): 1–40; Harsha Walia, *Undoing Border Imperialism* (Oakland, Calif.: AK Press, 2013).

7 Leslie Marmon Silko, *Almanac of the Dead* (New York: Penguin, 1994); Channette Romero, "Envisioning a 'Network of Tribal Coalitions': Leslie Marmon Silko's *Almanac of the Dead*," *American Indian Quarterly* 26, no. 4 (2002): 623–40.

8 For more on the idea of shifting into a new mode of thinking with regard to prior eras of colonialism, see Walter Mignolo, "Anomie, Resurgences, and De-Noming," in *The Anomie of the Earth: Philosophy, Politics, and Autonomy in Europe and the Americas*, ed. Federico Luisetti, John Pickles, and Wilson Kaiser (Durham, N.C.: Duke University Press, 2015), vii–xv.

9 Vine Deloria, Jr., "Traditional Technology," in *Spirit and Reason* (Boulder, Colo.: Fulcrum Press, 1999), 129–36.

10 Craig Howe, "New Architecture on Indigenous Lands" (presentation at the Illinois School of Architecture, Urbana, April 2014).

11 Craig Howe, "Cyberspace Is No Place for Tribalism," *Wicazo Sa Review* 13, no. 2 (1998): 19–28.

12 Subcomandante Insurgente Marcos, "The Hand That Dreams When It Writes," in *The Speed of Dreams: Selected Writings 2001–2007*, ed. Marco Canek Peña-Vargas and Greg Ruggiero (San Francisco: City Lights, 2006), 140–43.

13 Thomas L. Friedman. *The World Is Flat: A Brief History of the 21st Century* (New York: Farrar, Straus, and Giroux, 2005).

14 Miranda Belarde-Lewis, "Sharing the Private in Public: Indigenous Cultural Property in Online Media" (paper presented at the iConference, Seattle, February 2011).

15 Sweetwater Nannauck, interview by author, August 21, 2013.

16 Vine Deloria, Jr., "Traditional Technology," in *Spirit and Reason: A Vine Deloria, Jr., Reader* (Boulder, Colo.: Fulcrum Press, 1999), 129–36.

17 For examples of what happens with the rise of Indigenous elite classes striving for nationalism, see in particular Prasenjit Duara, *Decolonization: Perspectives from Now and Then* (New York: Routledge, 2004).

18 AbTec: Aboriginal Territories in Cyberspace, "About: Aboriginal Territories in Cyberspace Empowering First Nations with New Media Technologies," http://abtec.org/index.html (accessed September 3, 2015).

19 Upper One Games, "Never Alone / Kisima Innitchuna," http://neveralonegame.com (accessed September 3, 2015).

20 Paul Farrell, "Human Rights Groups Condemn Nauru's Criminalisation of Political Protest," *Guardian*, May 27, 2015, www.theguardian.com/world/2015/may/28/human-rights-groups-condemn-naurus-criminalisation-of-political-protest (accessed September 3, 2015); Deji Olokotun, "Why Is a Tiny Island Nation Facing an Internet Shutdown?" *Access: Mobilizing for Global Digital Freedom*, May 14, 2015, www.accessnow.org/blog/2015/05/14/why-is-a-tiny-island-nation-facing-an-internet-shutdown (accessed September 3, 2015).

CONCLUSION

1 Martin Heidegger, *The Question Concerning Technology and Other Essays*, ed. and trans. William Lovitt (New York: Harper Colophon, 1977); Vine Deloria, Jr., "Traditional Technology," in *Spirit and Reason: A Vine Deloria, Jr., Reader* (Boulder, Colo.: Fulcrum Press, 1999), 129–36.
2 Martin Heidegger, *The Question Concerning Technology and Other Essays*, ed. and trans. William Lovitt (New York: Harper Colophon, 1977).
3 Vine Deloria, Jr., "Traditional Technology," in *Spirit and Reason: A Vine Deloria, Jr., Reader* (Boulder, Colo.: Fulcrum Press, 1999).
4 United Nations General Assembly, "Article 20: The Promotion, Projection and Enjoyment of Human Rights on the Internet," Universal Declaration of Human Rights, 2012.
5 Vine Deloria, Jr., "The Right to Know" (paper presented at the US Department of the Interior, Office of Library and Information Services, Washington, D.C., 1978).

GLOSSARY

airspace: Refers to the territory over a sovereign country's political boundaries. In the United States, the Federal Aviation Administration manages the sovereign use of airspace. US federally recognized tribes do not have control of the airspace over tribal reservation lands. Airspace is not to be confused with airwaves.

airwaves: Shorthand term for radio waves, the portion of the earth's electromagnetic spectrum that ranges from 1 millimeter to 100 kilometers when measured in wavelengths. The Federal Communications Commission maps and licenses the rights to use particular frequencies of radio waves for different purposes, including television broadcasts, AM/FM radio, and wireless broadband communications. *See also* spectrum *and* unlicensed spectrum.

American Indian: A legal term that emerged out of common use by colonial authorities and settlers who, since the late 1500s, were erroneously describing the original indigenous inhabitants of what are now the Americas as *indios*, or "Indians." The term *American Indian* is used in many of the treaty documents negotiated between tribal peoples and US colonial authorities, even though tribal peoples continue to recognize themselves by the names of their tribes (i.e., Navajo or Diné, for Navajo Nation) and not according to the generalized population of indigenous peoples of the Americas or the English-language term *American Indian*, as neither articulates the inherent sovereign rights of tribes. In the 1960s, organizers of the American Indian Movement reclaimed the nomenclature as a source of intertribal, shared identity and empowerment among the indigenous, non-settler peoples of what is now the United States. US federal authorities continue to use the term *American Indian* and the updated term *Native American* to define indigenous peoples of what is now the United States as an exceptional class of minority citizens. Because of early treaty negotiations and the US policy of just treatment toward non-white socially disadvantaged citizens, American Indians and/or Native Americans are granted certain rights and support mechanisms, including federal funding for education and health-care services, in exchange for the wrongful, unjust, and ongoing claim of sovereign Native lands. For theoretical and scientific purposes, it is important to understand the term *American Indian* as a colonial tool for describing an indigenous US population in aggregate, regardless of the social and political distinctions of the many peoples of the United States. It is also important to recognize that American Indians are not ethnic minorities like Asian Americans, African Americans, and Hispanic Americans but are the modern descendants of the self-governing Indigenous peoples of what are now the Americas.

American Indian Movement (AIM): An organization that grew out of activism among young tribal leaders in the 1960s. AIM activism followed on the heels of 1950s civil rights era movements and coincided with separate and distinct Chicano (Raza), Second Wave feminist, and Black Panther organizing. Most notable for the 1970 occupation of Alcatraz and the television broadcast of US federal counterintelligence and blockade at Wounded Knee, AIM goals and values continue to inform Native and Indigenous modes of leadership, scholarship, and political organizing. These include an emphasis on strengthening understanding of US histories from intertribal American Indian perspectives, reclaiming sovereign tribal lands, freeing American Indian political prisoners, and sustaining a critique of US occupation and colonization of Native lands and lifeways. *See also* Indigenous.

Assembly of First Nations (AFN): An organization of the leaders of the inherently sovereign aboriginal peoples of what is now Canada. The AFN convenes for purposes of intertribal discussion, decision making, and organizing on matters affecting the sovereign rights of First Nations and the binding treaties between First Nations, Canada, and the United Kingdom. AFN goals somewhat parallel the goals of the National Congress of American Indians, although the mechanisms shaping the emergence and authority of these organizations differ significantly.

autonomy: In the context of Indigeneity, the word *autonomy* refers to the potential for the enactment of the political will of a self-governing Indigenous people. In the United States and Canada, this most often takes shape via the mechanisms of legal/political sovereignty; in Mexico, this takes shape via the mechanisms of the legal/political activism of autonomous Indigenous pueblos. For theoretical purposes, it is important to understand that in the context of Indigeneity, autonomy occurs via the free will of a people and not at the level of the individual. Group membership—meaning affiliation and established kinship with a spiritually distinct and land-based and/or linguistically unique people—is therefore an integral component of Indigenous autonomy.

bandwidth: Refers to the rate that information, measured in bits, is channeled through digital media devices, that is, two hundred megabytes per second of streaming broadband. With regard to wireless network systems, bandwidth is also measured in hertz, which refers to the spectrum frequency (*see also* spectrum). In this document, the word *bandwidth* is used in the context described by participants, which is most often with regard to needing devices and systems that can channel greater bandwidth for handling heavier streaming content, or more bytes per second. When participants describe the need for greater access to bandwidth in terms of hertz, this is expressed in terms of needing greater access to spectrum.

broadband: Generally speaking, refers to a digital communication channel of at least 256 kilobytes per second. Technically speaking, *broadband* refers to the ability of a device to transmit multiple signals across multiple channels: fiber-optic cable, coaxial cable, and wireless, for example. Since the late 1990s, broadband has come to refer colloquially to high-speed Internet, or Internet over 256 kilobytes per second and operating in distinct contrast to former modes of single-channel dial-up.

Broadband Initiatives Program (BIP): In 2009, President Obama signed the US Department of Agriculture American Recovery and Reinvestment Act (ARRA). Under this act, $2.5 billion were allocated to subsidize the build-out of broadband Internet infrastructures in rural communities. This subsidy program was named the Broadband Initiatives Program and was administered in three rounds, beginning in 2010, through the US Department of Agriculture's Rural Utility Service. Thus far, two rounds of funding have been administered. *See also* Broadband Technology Opportunities Program.

Broadband Technology Opportunities Program (BTOP): Under the 2009 American Recovery and Reinvestment Act, $4.7 billion were allocated to subsidize the build-out of broadband Internet infrastructures, public computing centers, and data gathering, with the funding for infrastructure and computing centers directed to unserved and underserved communities. The BTOP is administered through the National Telecommunications and Information Administration, with data gathering monitored by the Federal Communications Commission and made publicly available through the online publication of the National Broadband Plan and the National Broadband Map. *See also* Broadband Initiatives Program.

casinos: At present, casino gaming as a tribal enterprise represents a significant source of income for gaming tribes. For a number of reasons, not all tribes support gaming operations. Some tribes find gaming out of step with tribal spiritual practices. Others have chosen to invest in other forms of enterprise. Others have not focused on gaming in their strategic plans. Casinos are a source of controversy in Indian Country and at its borders. Almost annually, elected officials representing counties in the state of California—which has a unique relationship with the tribes—call for a tax on tribal gaming operations. What these officials fail to understand is that California already benefits significantly from having taken the lands and waters of California tribes. For many tribes, casinos are their only means of acquiring the capital needed to pay for basic social services to tribal members, such as health care, early childhood education, scholarships for adult learners, law enforcement equipment and personnel, paved roads, legal services, schools, libraries, and telecommunications and Internet infrastructure and services.

colonization: Colonization is the enactment of colonialism, which is at once a social policy and an expansionist ideology. Historically, colonialism has manifested in many different ways in many different communities, but at its core, it emerges as a set of relationships in which one social group continually and habitually profits by exploiting the living environments, bodies, social organization, and spiritualities of another social group. Colonialism is marked by generations of subjugation such that the profiting social group begins to build all its social structures and institutions around them in order to support both its belief in its superiority and its means of exploitative and violent profit making. While Native and Indigenous scholars are currently in an era of analyzing the pathways and mechanisms of colonialism and thus have not produced a sufficient account of how it works, Anibal Quijano's (2000) analysis of the coloniality of power is most useful for information science studies. In this book, I have operationalized colonization based on the following four overlapping mechanisms: (1) the classification of indigenous peoples as a single lower class of subhumans worthy of social subjugation at best and extermination at worst, (2) the theft and settlement of indigenous lands and social spaces by the so-called elite class, (3) the articulation of institutions that support this caste system and elite control of the environment, and (4) the disciplining of elite forms of knowledge through the marginalization of all indigenous languages, philosophies, spiritualities, and modes of self-government.

community: The word *community* is used rather loosely in this book but not without awareness of Benedict Anderson's (*Imagined Communities*) conceptualization of social groups being composed of not just face-to-face interactions but also shared beliefs about ancestors, hopes for future generations, and social customs and mores that bind them into a community, regardless of whether or not they interact in person each day. In tribes, there are many individuals who are members of the tribal community but are not enrolled tribal members, including, for example, non-tribal business owners, people who have married into the tribe, and neighbors who have developed deep ties with the people in that landscape. Generally speaking, when referring to community needs in this research, I am referring to the social groups and individuals who are perceived by technology project leaders as consumers or users of the project. Community information needs in tribes are also very much connected to the goals of cultural sovereignty. In writing about tribal communities, I am also making a tacit reference to the dimensions of peoplehood (Holm, Pearson, and Chavis, "Peoplehood"), in that tribes comprise an inherently sovereign people or confederated sovereign peoples who relate with one another every day on the basis of a long history of kinship, a shared indigenous language or languages, shared spirituality or ceremonial cycle, and a relationship with the landscape that goes back to time immemorial. In this way, a community of feminist booksellers (as described in Burnett, Besant, and Chatman, "Small Worlds") is significantly different from a tribal community.

constraints: Refers to those factors in an environment that shape the design, functionality, content, and uses of a system.

content: Refers to information that is channeled through a particular series of devices and designed to appear on a digital interface for human consumption. For example, the data of ones and zeros are channeled through fiber-optic cables and reassembled on a computer in the form of a digital black-and-white photo. The photo is the content, whereas the ones and zeros are a kind of information, specifically, a form of metadata.

cultural sovereignty: Refers to the reality of the existence of contemporary Native peoples as self-governing peoples free to live by the ways of knowing they have developed over millennia within the ecologies of their homeland. Thus cultural sovereignty relates to the ability of elders and experienced members to share ways of knowing with younger members. These ways of knowing pertain to the long history of kinship among the people, the people's indigenous language or languages, their spirituality or ceremonial cycle, and their ancient yet continuously unfolding sacred relationship with the landscape. Cultural sovereignty is often explained in comparison to legal/political sovereignty, which refers to the political rights contemporary Native peoples have negotiated with the federal government.

design: In this book, design is understood from a sociotechnical perspective laden with a bit of Vine Deloria, Jr.'s ("If You Think about It, You Will See That It Is True"), explanations of Native modes of creation. Here, the word *design* does not just describe the decisions project leaders make about workplace goals and objectives, functionality, usability, and the aesthetic appearance of a system; it also refers to the social policies, customs, habits and norms, and consideration of the landscape that shape individuals' decisions to build a system in the first place. In this way, the requirements of cultural sovereignty become part of the process for designing the layout and functionality of tribal broadband networks. This is best demonstrated in the design of the TDVnet backbone.

devices: Refers to the pieces of tangible hardware that constitute an information system. Devices are necessarily designed to operate with one another, and when they do not, humans modify devices or innovate new ones in order to encourage efficient system compatibility. A smartphone is a device, a fiber-optic cable is a device, a wireless modem is a device. A police car is a device. Even a pencil is a device. Devices are ultimately predicated on human use.

digital devices: Refers to pieces of tangible hardware that compose a digital information system, meaning a system built to function on the high-speed transmission and calculation of binary code. *See also* devices.

8(a) certification: In 1953, Congress passed the Small Business Act (15 U.S.C. 631) to promote opportunities for small and disadvantaged business to acquire loans and government contracts. Presently, the Small Business Administration hosts the 8(a) Business Development Program, in which businesses at least 51 percent owned and controlled by socially and economically disadvantaged individuals are eligible for business mentorship and enrollment as sole-source government contractors. Businesses that apply for 8(a) certification must prove feasibility, sustainability, and increases in revenues for a few years before receiving certification. In Indian Country, the Native Procurement Technical Assistance Program (Native PTAC) is a resource for Native-owned and tribal enterprises seeking 8(a) certification.

Federal Communications Commission (FCC): In 1934, Congress established the Federal Communications Commission as an independent government agency regulating interstate and international telecommunications. The FCC now regulates radio, television, wire, satellite, and cable communications and focuses on areas such as consumer protection, law enforcement, broadband, ICT innovation, spectrum regulation, and support for ICT enterprises.

FCC Office of Native Affairs and Policy: In August 2010, the Federal Communications Commission established the Office of Native Affairs and Policy, recognizing the need after input from workshops on telecommunications issues in Indian Country, notices of inquiry on connectivity issues in rural locations, and meetings with the FCC Native Broadband Task Force. Geoffrey Blackwell (Muscogee Creek) now heads the Office of Native Affairs and Policy. Since 2010, Blackwell and his team have been documenting issues with telecommunications and Internet service provision in Indian Country and promoting a number of policies and regulatory adjustments to ensure that tribes have access to federal subsidies and spectrum licensing.

federal recognition: The process of federal recognition is intriguing for Native and Indigenous scholars because, on the one hand, it is the point at which the US federal government acknowledges the original sovereign status of an indigenous people, meaning it acknowledges the primacy of the people's claim to the land; on the other hand, it does so by bestowing an array of legal rights that embed the people's mode of governance within the dominant US colonial apparatus. US federal authorities acknowledge that the United States must either hold in reserve or pay out to the original nation of people the original lands and other resources that it has been holding in trust for them because it has illegitimately occupied these lands. Furthermore, because the US federal government will continue to operate in a fashion that colonizes Native lands, waters, and bodies, those recognized as Native Americans within sovereign US borders have rightful access to various resources held in trust for them by the United States, as a kind of exchange for the illegitimate settlement of the land. Thus, it is with a strange irony that self-governing Native peoples of Turtle Island enter into the federal recognition process. On the one hand, they become eligible to receive a certain amount of resources that Native peoples need, in many cases, to survive under the contemporary unfavorable and oppressive colonial political arrangement. On the other hand, they also enter into an agreement to hold US federal authorities accountable for honoring the sovereign rights of tribes, including those not yet codified human rights that many tribes exercise through their own modes of self-governance, including the practice of customary laws and lifeways (Alfred, "Sovereignty"). Entering into a relationship of trust with the same colonial federal government and all its states, institutions, and citizens who normatively settle Native lands, waters, and bodies induces a unique conflict of interest for leaders of Native nations who must contribute to the lifeways of their people, while respecting the fundamentally colonial legal, political, and economic exercises of the dominant hegemonic government (Barker, *Sovereignty Matters*; Kauanui, *Hawaiian Blood*). It is perhaps for this reason that, after years of research, esteemed Indigenous studies and legal scholar Johnathon Goldberg-Hiller ("Borders of Kinship") regards the politics of recognition as inherently a politics of violence.

fiber-optic cable: Refers to a key broadband technology that allows for transmission of multiple kinds of information across multiple channels through a single medium. Known colloquially as "fiber," fiber-optic cable is made of bundled strands of glass or fabric sewn around a core and housed in a durable rubber coating. Fiber-optic cable represents one of the most durable means for transmitting broadband across great distances and can be run underground along roadsides (terrestrial deployment) or strung between telephone poles (aerial deployment).

fiber to the home: Shorthand reference to the method of deploying regional broadband by laying or stringing fiber-optic cable from anchor institutions to houses within a service community. Though the cable may be available for home hook-up, residents may still have to subscribe to broadband Internet service through the regional Internet service provider.

4G: Refers to the fourth generation of mobile phone telecommunications standards. 4G mobile phones support a range of functions, from mobile television to wireless modem, as well as applications designed to run on a WiMax network. Older mobile phones operate on 3G and 2G standards. 4G mobile phones work only in regions with system-compatible 4G and 3G WiMax network providers.

geopolitics: A useful analytic frame for understanding tribes as borderland societies, in which tribal borders and boundaries indicate more than just jurisdictional limitations and also refer to long-term, cyclical, and overlapping social policies and programs that reinforce a politics of social difference, conflict over land uses, and moral claims of ownership. Native and Indigenous studies historian Paige Raibmon ("Unmaking Native Space") described this in her analysis of the genealogies of Native land dispossession in the Pacific Northwest. Understanding the conflicts between Native and settler land claims from a geopolitical perspective also allows for the explanation of patterns of colonization and dispossession from a global historical economic perspective. Immanuel Wallerstein's *Politics of the World-Economy*) world-systems theoretical approach is quite reliable in this regard, as is Mignolo's (*Local Histories / Global Designs*) analysis of dispossession in the colonized Americas.

homelands: In this book, a tacit reference to the conceptualization of peoplehood as an inherently sovereign people or confederated sovereign peoples who relate with one another every day on the basis of a long history of kinship, a shared indigenous language or languages, shared spirituality or ceremonial cycle, and a relationship with the landscape that goes back to time immemorial (Holm, Pearson, and Chavis, "Peoplehood"). Therefore, it is not a homeland in a nationalist patriotic sense but rather as seen from a more Indigenous perspective, which finds homelands to be composed of a people's deeply ecological relationship with mountains, mesas, grasses, constellations, seasons, and animal beings. Many Native peoples bear allegiance on the basis not of a social contract of citizenship but of ancient sacred instructions and therefore an experience of cosmic belonging (Vine Deloria, Jr., "If You Think about It, You Will See That It Is True"; Nabakov, *Where Lightning Strikes*).

Indian Country: A legal term that refers to the federally recognized tribes and state-recognized tribes, pueblos, rancherias, bands, and Alaska Native villages and corporations within the political boundaries of the United States. Colloquially, the term also refers to Native peoples' habits and norms in this somewhat parallel society. As a legal term,

Indian Country has come to have meaning based on more than a century of treaty-making and recognition processes between Native peoples and US federal authorities. It inherently refers to an intertribal state of being for Native peoples in the United States.

Indigenous: There are two ways of approaching this term: as *indigenous* and as *Indigenous*. The term *indigenous* refers to the native species of a particular terrain. Early European explorers used this word to describe the native flora and fauna, including the people. Over time, an indigenous person came to be considered of a lower class. This is especially true in Latin American countries and is somewhat true in the United States, where, for many years, an Indian was considered a lesser person. Through many years of activism and international organizing, Native and Indigenous peoples around the world reclaimed the derogatory word *indigenous* and replaced it with the term *Indigenous*, which refers to individuals who are witnessing the rise of a global consciousness of the state of political exigency of Native and Indigenous peoples, specifically with regard to colonization as an abettor of the industrial and technological advancement of modern nation-states. In this way, Indigeneity represents a deeply modern and global mode of social critique. Indigenous scholars are trained in two ways: by respecting and learning from their own peoples' tribal and spiritual histories and by leveraging the tools of Western academia toward staging a strategic critique of nation-state colonial policies. Because of the global, politicized nature of Indigeneity, many people in tribal communities do not use this term to describe local spiritual activities or daily work practices. They do this out of respect for the homelands and the beings therein; the stridency and political urgency of Indigeneity are sometimes out of step with local ceremonial and spiritual obligations.

information: Refers to any item, tangible or intangible, that can function as a form of currency that charges and reproduces loosely bound networks of humans and devices. This is something of a break from the information scientific hierarchy in which meta-data builds up to data, which builds up to information, which builds up to knowledge. To an Indigenous information scientist, there is no objective measurable distinction between data, information, and knowledge, because Native peoples have been considered lacking in ways of knowing for centuries. Whole cultures have been broken apart and treated as pieces of information available for exploitation as a form of currency in world markets. One society's information is another's sacred ways of knowing. Data are to a human as electrical power is to a machine. It is, however, possible to theorize that in an imagined network society—in which humans and devices continually interrelate and reproduce institutions, technical devices, social policies, and norms—any item that is useful to this reproduction can be somehow broken down into a bit (pun intended) of useful information.

information and communication technologies (ICTs): Refers broadly to digital devices that allow for the synchronous and asynchronous exchange of digital content between humans. The term encompasses smartphones, laptops, tablets, mobile phones, streaming radio, and so forth.

information sharing: Refers to the intentional human-to-human exchange of information and knowledge, which can occur face-to-face or via ICT media. The term draws attention to the relational aspect of information: it is meaningful only when it is shared.

information system: Refers to an intentional assemblage of humans, devices, and policies designed to guide the flow of specific kinds of information to support workplace goals and objectives. Some information systems are small and localized, such as an office system for handling internal mail. Some are much larger in scale, such as the Red Spectrum Communications fiber-to-the-home network. The design of information systems in tribal administrative workplaces is shaped by policies as complex as the Indian Education and Self-Determination Act of 1975.

inherent sovereignty: Refers to the will of a people to self-govern, regardless of the official recognition of an overarching federal or colonial authority.

institution: In this book, refers to the ways that sociologists have conceptualized certain social structures—organizations, social practices, ideologies, social classes—that have crystallized in the form of governing useful socially reproductive and disciplinary bureaucracies. Universities, hospitals, and prisons are institutions. For contemporary Native peoples, the federally recognized tribal administration represents an institution, a concept with which inherently sovereign peoples tangle as they go through the paperwork and politicking required for the federal recognition process. Tribal colleges and the National Congress of American Indians are also institutions.

knowledge: Hawaiian studies scholar Manu Aluli Meyer (*Hoʻoulu*) has pointed out the distinction between knowing and knowledge, in that ways of knowing represent the motion of Indigenous elders and leaders imparting the keys to understanding philosophies and spiritualities alongside youth, and knowledge represents the observable products of this motion. Ways of knowing and knowledge both bear qualities of information that has been agreed upon as legitimate or "true" many times over, possess a reverential or experiential aspect, and are worthy of being recorded, codified, or passed on for purposes of institutional continuation, moral rectitude, or the survival of a people.

landscapes: In this book, refers to ecological environments imbued with human histories and experiences.

leapfrogging: Refers to the phenomenon by which certain communities have advanced technologically by skipping a stage of infrastructural capacity building. For example, some of the tribes in the Southern California Tribal Chairmen's Association leapfrogged when they acquired broadband Internet services before landline phone lines were in place.

legal/political sovereignty: At present, federally recognized tribes within the boundaries of the United States exercise the following eight rights as sovereign governments: the rights to self-govern, determine citizenship, and administer justice; the rights to regulate domestic relations, property inheritance, taxation, and the conduct of federal employees; and the right to sovereign immunity. These rights have been negotiated over more than a century of treaty making, court cases, and entanglements with the federal recognition process.

local area network: Refers to a system of cables and wireless connections set up to distribute Internet services to a number of nearby workstations, usually within a single office building.

National Congress of American Indians (NCAI): In 1944, a number of tribal leaders, lawyers, and American Indian studies scholars met in Denver, Colorado, to found a nonprofit organization dedicated to advocating for the rights of Native peoples in the United States. The NCAI hosts two meetings per year, in which the leaders of Indian Country share ideas and prepare to advocate for policy changes before the US Congress and within their tribal communities.

Native American: The latest term designated by the US federal government to refer to the exceptional class of US citizens descended from the original Indigenous peoples of what is now the United States. It replaced the term *American Indian* in the mid-1990s, when the United States began to adopt the social policy of multiculturalism. Unfortunately, in part because of this change in terminology, many people mistake Native Americans for ethnic minorities, as the term is similar to those designating other ethnic minorities, such as *African American* and *Asian American*. *See also* American Indian.

Native and Indigenous: A phrase that Indigenous studies scholars began using in the past decade or so to refer to the articulation of American Indian studies alongside global Indigenous studies.

network: Used in three different ways in this book. The first refers to the assemblage of humans, devices, and policies that make broadband infrastructures function in useful ways. Because broadband infrastructures are digital, network systems are digital information systems dealing with broadband technologies (i.e., spectrum, fiber-optic cables, wireless dishes, etc.). The second is in reference to Manuel Castells's (*The Information Age*) description of a network society, in which networked ICTs are enabling the global articulation of local social movements, work patterns, government policies, and the like. Underlying all of this, I have also relied on a third use of the term *network*, as outlined according to actor network theory, which holds that sociologists can understand the impacts of technologies by analyzing all technologies as assemblages of human technicians and nonhuman technical devices oriented toward a purposive work goal. In this way, contemporary society—a networked society, Castells might say—is really sociotechnical, as there is no longer any real division between man and machine. As such, a technical network system is related to a networked society, but their relationships in terms of scale and meaning must be localized to the context of the technical system's reason for being. The Coeur d'Alene Tribe's network backbone exists for the purpose of cultural sovereignty. Incorporated as Red Spectrum Communications ("Red" referring to the symbolic color of the American Indian Movement and "Spectrum" referring to the airwaves over tribal lands), the Coeur d'Alene network backbone as a technical system has been positioned by its project leaders as a kind of political statement within the global dynamic shaping American Indian cultural sovereignty.

reframing: Refers to a decolonizing methodology through which a social problem often diagnosed as an "Indian problem" is subverted to show that it is actually an outcome of overlapping patterns of colonization (Smith, *Decolonizing Methodologies*). Reframing means learning about the history of an "Indian problem" and making decisions about what to foreground in order to reveal the conditions shaping the colonial misdiagnosis of a social problem as the fault of Indigenous peoples and their ways of life. It is a decolonizing methodology in that the act of rewriting the history allows the writer not only to learn a new, more hopeful perspective on a complex social issue but also to surface keys to more durable long-term solutions. These solutions usually must emerge from the Indigenous communities themselves. Reframing is also a decolonizing methodology because it encourages readers to rethink their positions on the subject and see the problem from an Indigenous perspective. As a decolonizing methodology, its purpose is to reveal and subvert colonizing logics while creating the conditions for imagining Indigenous community-based solutions.

reservation communities: The difference between reservation communities and tribal communities is a fine but important distinction in Indian Country. Tribal communities consist of the members of a tribe, inclusive of their relationships. Tribal governments form on the basis of serving tribal communities, regardless of where tribal members reside. Reservation communities, however, consist of all the individuals who reside on or near the reservation, regardless of their affiliation with the tribe. While these can feel like one and the same on occasion, people who have grown up on or near reservations understand all too well the fine distinction between those who are working to support the tribal community around the basis of cultural sovereignty and those who profit from the violence that occurs at reservation borders, such as the liquor store operators, drug dealers, and other exploitative sorts who plague reservation communities. Sometimes they can include tribal relatives, family members, and friends. A tribal leader is more likely to speak of serving a tribal community.

reservation system: In this book, referred to as a system because of the rigorous method by which all Native peoples of the United States were coerced into this socially detrimental manner of living on restricted acreage out of step with their natural manner of living and moving across the homelands. The reservation system was introduced through hundreds of treaty documents, federal recognition agreements, and federal Indian policies over the past century or more. Reservations were, almost as a rule, placed on what federal authorities perceived as untamable, unusable, or polluted land, with military forts on-site or nearby. Winona LaDuke (*All Our Relations*) published a map of reservations collocated with US mining and nuclear power operations. Even those tribes that have managed to revitalize and create healthy homelands within reservation boundaries experience the stifling conditions of reservation life under the coercive laws and policies enforced by the US federal government and courts.

right relation: To "live in right relation" or "for all my relations" is a phrase often heard in Indian Country. It stems from an understanding that all beings are connected in a cosmic dynamic, including those who have passed and those yet to be born. To live in right relation is to live in a state of humility and with deference to others and to the living landscape. In this sense, humankind definitely is not at the forefront of a hierarchy of species evolution but is actually no more meaningful than a dust mite or a sea otter. To live in right relation is to recognize one's place in an ecology of beings, to have respect for the purpose of beings other than oneself and cycles other than one's own.

self-determination: A shorthand reference to the social policy underpinning the 1975 Indian Education and Self-Determination Act. Before this act, social services and other programs on reservations were designed and administered by federal personnel, such as employees of the Bureau of Indian Affairs. Social services and basic infrastructural programs were poorly designed. Federal authorities had no interest in strengthening Native communities. *Cobell v. Salazar* demonstrated the degree to which federal authorities had stolen funds and other resources from tribes. After passage of the Indian Education and Self-Determination Act, tribal peoples could create and implement their own programs. We are at present still in the era of self-determination, in which the federal government is charged with consulting with, supporting, and assisting tribes as necessary but is not to intervene.

self-governance: Realization of the difference between precolonial Indigenous modes of self-governance and the colonial form of government is essential to understanding self-governance. To retain the memory of these precolonial modes, a free and autonomous Native people will share information among themselves and with neighbors in order to strengthen their knowledge of their homeland, shared history, Native language, ceremonial cycle, and lineage. A particular spirituality and a particular philosophy of self-governance emerge out of these centuries of sharing ways of knowing. While the term *self-governance* refers to the cultural sovereignty of a people, meaning precolonial modes of governance and social organization, it also refers to modes of self-governance in the context of domestic dependency within the United States. The present colonial arrangement is based on the policy of self-determination: federally recognized Native peoples should and can determine the course of their own social services and civic arrangements within the federally recognized boundaries of their land. Decision making by tribal leaders on designing and managing domestic relations is a matter of self-governance.

settler colonialism: Pertains to the form of colonialism in which a dominant world-governing power claims territory in Indigenous territory and then sends its citizen-subjects to complete the colonization processes of settlement and expansion, ultimately displacing or attempting to assimilate the Indigenous peoples in the name of am imagined new form of nationalism. Australia, Canada, New Zealand, and the United States are settler colonial governments, with Britain representing the initiating colonial government. Settlers, or non-Indigenous subjects of a settler colonial government, in these countries eventually realized their own form of government, theoretically free of ties to their mother country; however, their new forms of government continued to depend on the eradication of the Native and Indigenous peoples of these territories. The habitus of settlers tends to be totalizing, in that their sense of nationalist belonging depends on their constant efforts to erase the languages, philosophies, legal claims to territory, scientific logics and practices, social norms, physical bodies, gender norms, and naturally occurring biomes of the original Indigenous peoples. From an Indigenous perspective, settler colonial narratives and logics appear delusional, illusory, self-referential, and designed to reify the expansionist economic and political practices of settler colonial institutions.

sociotechnical: According to actor network theory, sociologists can understand the impacts of technologies by analyzing all technologies as assemblages of human technicians and nonhuman technical devices oriented toward a purposive work goal. In this light, contemporary society is really sociotechnical, as there is no longer any real division between man and machine, which have become quite intertwined in contemporary society. This is even more interesting from an Indigenous perspective, which, generally speaking, does not find a real difference between man, or nature, and machine, or technology. Since humans are of nature, and it is within their disposition to create, they will create out of nature for nature. Actor network theory is quite helpful in this regard, in that the analytical method is based on understanding interactions between human and nonhuman actors, which might be a useful analytic for teaching non-Indigenous thinkers about human and animal relationships, human and mountain relationships, human and star-being relationships, and so on. Though it is useful in teaching thinkers about relationships between humans and technical devices—revealing that technical devices are deeply human in design and purpose and also that humans are capable of behaving systemically—it also strangely reifies the differences between humans and the devices they create.

spectrum: Refers to a measure of the wavelengths across which communications signals travel. The Federal Communications Commission distributes licenses for particular frequencies for broadcasting radio, television, broadband satellite, and wireless signals. *See also* airwaves *and* unlicensed spectrum.

state recognition: Refers to a process by which tribes are recognized by state authorities as Native peoples. It is somewhat parallel to the federal recognition process but quite different in terms of the resulting legal rights. Many California tribes are state-recognized but not federally recognized, as are some of the tribes receiving Internet service through TDVnet. *See also* federal recognition.

stories: For Native and Indigenous scholars, stories represent a particular kind of incontrovertible evidence that is not to be triangulated but rather folded into a broader understanding of complex and multi-perspectival phenomena. Stories are accepted as a medium for passing on knowledge, building an interpersonal relationship, providing witness to an event, guiding right thinking on a matter, and expressing a sensation or experience.

system: In this book, the word *system* is used in two different ways: information systems and social systems. Information systems are intentional assemblages of humans, devices, and policies designed to guide the flow of specific kinds of information to support workplace goals and objectives. Social systems (e.g., the reservation system) can be understood as a series of interoperable practices and norms guiding the disciplinary work of institutions. This work also makes tacit reference to Immanuel Wallerstein's (*Politics of the World-Economy*) world-systems analysis, in which geopolitical societies can be broken into cores and peripheries that exist in an economically capital-productive, yet inherently violent, power dynamic. Understanding pathways and mechanisms of colonization means using a systems approach to understand the economic reasons for the global circulation of people, goods, and currency.

technology: In more general discussions, this book favors the more specific terms *devices*, *digital devices*, and *information and communication technologies*, or *ICTs*, rather than the word *technology*. On technology, Vine Deloria, Jr. ("Traditional Technology"), reads to the Greek root of the word, *techne*, to arrive at the significance of a human bringing a tool into being. Through its making, the tool becomes imbued with the goals and intentions of the designer. Martin Heidegger (*Question Concerning Technology*) refers to the functionality and values embedded in the creation of a silver chalice. Deloria refers to the methods, or techniques, by which spiritual people were able to bring into being the sensation of cosmic order and purpose and reminds Native students that their purpose in designing technologies is to restore the health of Native bodies, minds, and homelands. *See also* sociotechnical.

tribal council: Generally speaking, refers to a contemporary mode of self-governance in Indian Country, in which tribal government leadership is shared among a number of elected officials who serve for limited terms. Tribal council members are most often positioned to act as liaisons, idea people, entrepreneurs, and policy advocates between the tribal community and federal government agencies. Their difficult work is never done.

unlicensed spectrum: While the Federal Communications Commission administers licenses for some frequencies, others remain unlicensed for purposes of innovation and emergency uses. TDVnet now runs almost entirely on unlicensed spectrum. As designated spectrum licenses can be quite expensive, and some Internet service providers have been known to squat on spectrum, holding on to it but never using it, the FCC is looking into creating a tribal priority for licensed spectrum, a use-it-or-lose-it policy, and opening up frequencies for more unlicensed uses. *See also* airwaves *and* spectrum.

ways of knowing: Articulates the dynamic nature of Native philosophies, histories, and land-based spiritualities. The term *ways of knowing* is used in counterpoint to the static *knowledge*. This is done partly in reaction to depictions of Native peoples in museums and libraries as dead and gone, objects of knowledge, rather than as living and continually creating ways of knowing. *See also* knowledge.

BIBLIOGRAPHY

AbTec: Aboriginal Territories in Cyberspace. "About: Aboriginal Territories in Cyberspace Empowering First Nations with New Media Technologies." Accessed September 3, 2015. http://abtec.org/index.html.

Adas, Michael. *Machines as the Measure of Men: Science, Technology, and Ideologies of Western Dominance*. Ithaca, N.Y.: Cornell University Press, 1989.

Agrawal, Arun. "Indigenous Knowledge and the Politics of Classification." *International Social Sciences Journal* 54, no. 173 (2002): 287–97.

Akins, Damon. "Pioneer Nostalgia, the Spanish Fantasy Past, and the Emergence of the Mission Indian Federation, 1870–1920." Paper presented at the Native American Indigenous Studies Association, Sacramento, California, June 2011.

Alfred, Taiake. "Sovereignty." In *Sovereignty Matters: Locations of Contestation and Possibility in Indigenous Struggles for Self-Determination*, edited by Joanne Barker, 33–50. Lincoln: University of Nebraska Press, 2005.

Anderson, Benedict. *Imagined Communities: Reflections on the Origin and Spread of Nationalism*. New York: Verso, 1983.

Bang, Megin, and Douglas Medin. *Who's Asking? Native Science, Western Science, and Science Education*. Cambridge, Mass.: MIT Press, 2014.

Barehand Green, Shana. "Native and On-Reservation Business by Non-Natives." Lecture presented at the Native Procurement and Technical Assistance Center Native American and Veteran Small Business Conference and Tradeshow, Tulalip, Washington, May 2012.

Barker, Joanne, ed. *Sovereignty Matters: Locations of Contestation and Possibility in Indigenous Struggles for Self-Determination*. Lincoln: University of Nebraska Press, 2005.

Bauer, William J., Jr. "When the Owens Valley Went Dry: Politics, Water, and the Paiute Oral Tradition in the 1930s." Paper presented to the Native American and Indigenous Studies Association, Sacramento, California, June 2011.

Beaton, Brian. "Online Resources about Keewaytinook Okimakanak, the Kuhkenah Network (K-Net) and Associated Broadband Applications." *Community Informatics* 5, no. 2 (2009). http://ci-journal.net/index.php/ciej/article/view/571/448.

Belarde-Lewis, Miranda. "Sharing the Private in Public: Indigenous Cultural Property in Online Media." Paper presented at the iConference, Seattle, February 2011.

Bijker, Wiebe, and John Law. *Shaping Technology/Building Society: Studies in Sociotechnical Change*. Cambridge, Mass.: MIT Press, 1992.

Bissell, Theresa. "The Digital Divide Dilemma: Preserving Native American Culture While Increasing Access to Information Technology on Reservations." *Journal of Law, Technology, and Policy* 1 (2004): 129–50.

Bock, Joseph. *The Technology of Nonviolence: Social Media and Violence Prevention*. Cambridge, Mass.: MIT Press, 2012.

Bowden, Charles. *Blue Desert*. Tucson: University of Arizona Press, 1986.

———. *Murder City: Ciudad Juarez and the Global Economy's New Killing Fields*. New York: Nation Books, 2010.

Bowers, Chet, Miguel Vasquez, and Mary Roaf. "Native Peoples and the Challenge of Computers: Reservation Schools, Individualism, and Consumerism." *American Indian Quarterly* 24, no. 2 (2000): 182–99.

Bowker, Geoffrey, and Susan Leigh Star. *Sorting Things Out: Classification and Its Consequences.* Cambridge, Mass.: MIT Press, 1999.

Bowman, Christopher. "Indian Trust Fund: Resolution and Proposed Reformation to the Mismanagement Problems Associated with the Individual Indian Money Accounts in Light of *Cobell v. Norton.*" *Catholic University Law Review* 53, Cath. U. L. Rev. 543 (2004): 543–1195.

boyd, danah. *It's Complicated: The Social Lives of Networked Teens.* New Haven, Conn.: Yale University Press, 2014.

Brave Heart, Maria Yellow Horse. "Models for Healing, Indigenous Survivors of Historical Trauma, Theory and Research Implications for the Seattle Community." Lecture presented at the Indigenous Wellness Research Institute, University of Washington, Seattle, October 2009.

Buddle, Kathleen. "Aboriginal Cultural Capital Creation and Radio Production in Urban Ontario." *Canadian Journal of Communication* 30, no. 1 (2005): 7–39.

Burnett, Gary, Michele Besant, and Elfreda Chatman. "Small Worlds: Normative Behavior in Virtual Communities and Feminist Bookselling." *Journal of the American Society for Information Science and Technology* 52, no. 7 (2001): 536–47.

Busacca, Jeremy. "Seeking Self-Determination: Framing, the American Indian Movement, and American Indian Media." PhD diss., Claremont Graduate University, 2007.

Cajete, Gregory. *Native Science: Natural Law of Interdependence.* Santa Fe, N.M.: Clear Light Publishers, 2000.

Canclini, Néstor García. *Imagined Globalization.* Durham, N.C.: Duke University Press, 2014.

Carlson, W. Bernard. *Tesla: Inventor of the Electrical Age.* Princeton, N.J.: Princeton University Press, 2013.

Casey, James, Randy Ross, and Marcia Warren. *Native Networking: Telecommunications and Information Technology in Indian Country.* Washington, D.C.: Benton Foundation, 1999.

Castells, Manuel. *The End of Millennium.* Vol. 3 of *The Information Age: Economy, Society and Culture*. Malden, Mass.: Blackwell Press, 1998.

———. "A Network Theory of Power." *International Journal of Communication* 5 (2011): 773–87.

———. *The Power of Identity.* Vol. 2 of *The Information Age: Economy, Society and Culture*. Malden, Mass.: Blackwell Press, 1997.

———. *The Rise of the Network Society.* Vol. 1 of *The Information Age: Economy, Society and Culture*. Malden, Mass.: Blackwell Press, 1996.

Caven, Febna. "Being Idle No More: The Women behind the Movement." *Cultural Survival Quarterly* 37, no. 1 (2013): 6–7.

Cernera, Phillip, Anne Dailey, and Rebecca Stevens. "Superfund and NRDA Working Together: Mine Waste to Healthy Habitat." Lecture presented at the Institute for Tribal Environmental Professionals Tribal Lands and Environment Forum, Flagstaff, Arizona, 2012.

Chan, Anita Say. *Networking Peripheries: Technological Futures and the Myth of Digital Universalism*. Cambridge, Mass.: MIT Press, 2014.

Chapin, Mac, Zachary Lamb, and Bill Threlkeld. "About Us." Cheyenne River Sioux Tribe Telephone Authority. Accessed January 10, 2013. http://www.crstta.com/about-us.

———. "Mapping Indigenous Lands." *Annual Review of Anthropology* 34 (2005): 619–38.

Cheyenne River Sioux Tribe Telephone Authority. *Before the Federal Communications Commission in the Matter of Petition of the Cheyenne River Sioux Tribe Telephone Authority for*

Designation as an Eligible Telecommunications Carrier Pursuant to Section 214(e)(6) of the Telecommunications Act of 1996. FCC 97-419 (1998).

———. Comments before the Federal Communications Commission. *In the Matter of Inquiry regarding Current Carrier Systems, Including Broadband over Power Lines Systems, April 29, 2004, Amendment of Part 15 regarding New Requirements and Measurement Guidelines for Access: Broadband over Power Line Systems, April 29, 2004 (ET Docket No. 04-37)*. Accessed November 5, 2016. Available at https://ecfsapi.fcc.gov/file/6516182369.pdf.

Cheyenne River Sioux Tribe Telephone Authority and US West Communications, Inc. *Appellants v. Public Utilities Commission of South Dakota*, Appellee (1999) nos. 20062, 20464.

Clarkson, Gavin, Trond Jacobsen, and Archer Batcheller. "Information Asymmetries and Information Sharing." *Government Information Quarterly* 24 (2007): 827–39.

Cleaver, Harry. "The Zapatista Effect: The Internet and the Rise of an Alternative Political Fabric." *Journal of International Affairs* 51, no. 2 (1998): 620–40.

Clegg, Stewart. *Frameworks of Power*. London: Sage, 1989.

Clements, William. *Imagining Geronimo: An Apache Icon in Popular Culture*. Albuquerque: University of New Mexico Press, 2013.

Clinton, William Jefferson. "Executive Order 13175: Consultation and Coordination with Indian Tribal Governments." *Federal Register* 65, no. 218 (2000): 67249–51.

———. "Remarks to the People of the Navajo Nation in Shiprock, New Mexico, April 17, 2000." In *Public Papers of the Presidents*, edited by John Woolley and Gerhard Peters. The American Presidency Project. Accessed March 8, 2013. www.presidency.ucsb.edu/ws/index.php?pid=58134.

Cobb, Amanda. "Understanding Tribal Sovereignty: Definitions, Conceptualizations, and Interpretations." *American Studies* 46, nos. 3/4 (2005): 115–32.

Coeur d'Alene Tribe. "About Sovereignty." Accessed January 10, 2013. www.cdatribe-nsn.gov/cultural/sovereignty.aspx.

Committee on Superfund Site Assessment and Remediation in the Coeur d'Alene River Basin and the National Research Council. *Superfund and Mining Megasites: Lessons from the Coeur d'Alene River Basin*. Washington, D.C.: National Academies Press, 2005.

Computerworld Honors Program. "Department of Interior, Navajo Nation: Internet to the Hogan Project." 2007. Accessed January 11, 2013. www.cwhonors.org/viewCaseStudy.asp?NominationID=250.

Conner, Debbie. "'To the Most Exacting Fiduciary Standards': Notes on Information Technologies and Federal Administration of Indian Trust Lands and Resources." *Wicazo Sa Review* 13, no. 2 (1998): 99–116.

Copps, Michael. "Welcoming Address." Lecture presented at the Tribal Telecom and Technology Summit, Gila River, Arizona, 2012.

Cordero, Carlos. "Reviving Native Technologies." In *Surviving in Two Worlds: Contemporary Native American Voices*, edited by Louis Crozier-Hogle, Daryll Babe Wilson, and Jaye Leibold, 83–92. Austin: University of Texas Press, 1997.

Coyne, Richard. *The Tuning of Place: Sociable Spaces and Pervasive Digital Media*. Cambridge, Mass.: MIT Press, 2010.

Croffut, George. *Crofutt's New Overland Tourist and Pacific Coast Guide*. Omaha, Neb.: Overland, 1878.

Curran, John. "Governance, Communications Technologies, and the Transition to IPV6." Presentation at the Tribal Telecom and Technology Summit, Gila River, Arizona, February 2014.

Cutright, Erik. "Tribal Telecom Enterprises and ISPs: Connecting ALL the Dots: Karuk Tribe." Lecture presented at TribalNet, San Diego, California, 2012.

Dampier, Phillip. "Montana's Struggle for Broadband Pits Cable, Phone Companies, and Native American Communities against One Another." Stop the Cap! Promoting Better Broadband, Fighting Data Caps, Usage-Based Billing, and Other Internet Overcharging Schemes. February 10, 2010. Accessed March 9, 2013. http://stopthecap.com/2010/02/10/montanas-struggle-for-broadband-pits-cable-phone-companies-and-native-american-communities-against-one-another/.

Dawes Act, or General Allotment Act of 1887, 25 § 331.

DeBruyn, Hans. *The First Nations ISP Guide: Providing Internet Services, Managing Operations.* West Vancouver, British Columbia, Canada: First Nations Technology Council, 2012.

de Landa, Manuel. *A Thousand Years of Nonlinear History.* Cambridge, Mass.: MIT Press, 2000.

De León, Jason, and Mike Wilson. "Rights, Sovereignty, and Lives on the Line: Immigration Debates across Arizona and Tohono O'odham Borderlands." Lecture presented at the Center for Global Studies, University of Washington, Seattle, February 2010.

Deloria, Phil. *Indians in Unexpected Places.* Lawrence: University of Kansas Press, 2004.

Deloria, Vine, Jr. "If You Think about It, You Will See That It Is True." In *Spirit and Reason: A Vine Deloria, Jr., Reader*, 40–62. Boulder, Colo.: Fulcrum Press, 1999.

———. "The Right to Know." Paper presented at the US Department of the Interior, Office of Library and Information Services, Washington, D.C., 1978.

———. "Traditional Technology." In *Spirit and Reason: A Vine Deloria, Jr., Reader*, 129–36. Boulder, Colo.: Fulcrum Press, 1999.

Deloria, Vine, Jr., and Clifford Lytle. *American Indians, American Justice.* Austin: University of Texas Press, 1983.

Digital Arizona Council. *Arizona's Strategic Plan for Digital Capacity: Expanded and Reference Version.* Phoenix: Arizona Department of Administration, Arizona Strategic Enterprise Technology, 2012.

Dorr, Jessica, and Richard Akeroyd. "New Mexico Tribal Libraries: Bridging the Digital Divide." *Computers in Libraries* 21, no. 8 (2001): 8.

Dourish, Paul, and Genevieve Bell. *Divining a Digital Future: Mess and Mythology in Ubiquitous Computing.* Cambridge, Mass.: MIT Press, 2011.

Drinnon, Richard. *Facing West: The Metaphysics of Indian-Hating and Empire-Building.* Minneapolis: University of Minnesota Press, 1980.

Duara, Prasenjit. *Decolonization: Perspectives from Now and Then.* New York: Routledge, 2004.

Duarte, Marisa Elena. "Connected Activism: Indigenous Uses of Social Media for Shaping Political Change." Paper presented at "By the People: Participatory Democracy, Civic Engagement, and Citizenship Education," Tempe, Arizona, December 3, 2015.

———. "Knowledge, Technology, and the Pragmatic Dimensions of Self-Determination." In *Restoring Indigenous Self-Determination*, edited by Marc Woons, 45–56. Bristol, South West England: E-International Relations, 2014.

Duarte, Marisa Elena, Miranda Belarde-Lewis, and Allison B. Krebs. "Native Systems of Knowledge: Indigenous Methodologies in Information Science." Lecture presented at the iConference, Urbana, Illinois, February 2010.

Dyck, Noel. *What Is the Indian Problem? Tutelage and Resistance in Canadian Indian Administration.* St. Johns, Newfoundland, Canada: Institute of Social and Economic Research, 1991.

Dyson, Laurel Evelyn, Max Hendriks, and Stephen Grant. *Information Technology and Indigenous People.* Hershey, Penn.: Information Science Publishing, 2007.

Ellul, Jacques. *The Technological Society.* New York: Vintage Books, 1964.

———. *The Technological System.* New York: Continuum, 1980.

Ereaux, Jim. "The Impact of Technology on Salish Kootenai College." *Wicazo Sa Review* 13, no. 2 (1998): 117–35.

Estrada, Gabriel. "Native Avatars, Online Hubs, and Urban Indian Literature." *Studies in American Indian Literatures* 23, no. 2 (2011): 48–70.

Evans, Randy, and Jim Dunstan. *Communications Regulation and Taxation in Indian Country: Executive Summary: A Report Prepared in Conjunction with the Tribal Telecom 2013 Conference.* Law Office of Randal T. Evans, PLLC, Phoenix, Arizona, and Mobius Legal Group, PLLC, Springfield, Virginia, 2013.

———. *Tribal Broadband Guide: Telecommunications Regulation and Taxation in Indian Country.* Washington, D.C.: Native American Finance Officers Association, 2013.

Fanon, Franz. "Concerning Violence." In *The Wretched of the Earth*, 35–106. New York: Grove Press, 1963.

———. *The Wretched of the Earth.* New York: Grove Press, 1963.

Farrell, Paul. "Human Rights Groups Condemn Nauru's Criminalisation of Political Protest." *Guardian*, May 27, 2015. Accessed September 3, 2015. www.theguardian.com/world/2015/may/28/human-rights-groups-condemn-naurus-criminalisation-of-political-protest.

Federal Communications Commission. "Connecting America: National Broadband Plan, 2010." Accessed January 15, 2011. www.broadband.gov/plan.

———. "FCC Chairman Genachowski Appoints Geoffrey Blackwell to Lead New Initiatives on Native Affairs." *FCC News*, June 22, 2010. Accessed June 13, 2013. http://hraunfoss.fcc.gov/edocs_public/attachmatch/DOC-298924A1.pdf.

———. *Federal Communications Commission Office of Native Affairs and Policy 2012 Annual Report.* Washington, D.C.: FCC, 2012.

———. "Tribal Consultation and Coordination Priorities for 2013." In *Office of Native Affairs and Policy, Consumer and Governmental Affairs Bureau, Federal Communications Commission Annual Report.* Washington, D.C.: FCC, 2012.

Federal Power Commission v. Tuscarora Indian Nation 362 U.S. 99 (1960).

Federici, Silvia. "Re-enchanting the World: Technology, the Body, and the Construction of the Commons." In *The Anomie of the Earth: Philosophy, Politics, and Autonomy in Europe and the Americas*, edited by Federico Luisetti, John Pickles, and Wilson Kaiser, 202–14. Durham, N.C.: Duke University Press, 2015.

Field, Tom. "Homegrown Talent: One Sioux Reservation Is Investing Its Future in an IT Services Venture Called Lakota Technologies, Inc." *CIO Magazine*, October 1, 2001. Accessed January 10, 2013. www.cio.com/article/30568/Lakota_Technologies_and_American_Indian_IT_Outsourcing.

Findlay, John. "Historic Settlers and Native Americans." In *A Cultural Resources Overview and Inventory of the Proposed Thomas-Newville Reservoir, Glenn and Tehama Counties, California: A Report Prepared for the California Department of Water Resources, Northern District, Red Bluff, California*, edited by Basin Research Associates, Inc., and Cultural Systems Research, Inc., 37–55. San Leandro, Calif.: Basin Research Associates, 1983.

Frank, Ross. "The Tribal Digital Village: Technology, Sovereignty, and Collaboration in Indian Southern California." Unpublished manuscript, University of California, San Diego, 2004. http://pages.ucsd.edu/~rfrank/class_web/ETHN200C/TDVchapters.pdf.

Fricker, Miranda. *Epistemic Injustice: Power and the Ethics of Knowing.* Cambridge: Oxford University Press, 2007.

Friedman, Thomas. *The World Is Flat: A Brief History of the 21st Century.* New York: Farrar, Straus, and Giroux, 2005.

Fuchs, Christian. *Social Media: A Critical Introduction.* London: Sage, 2014.

Garrido, Maria. "The Zapatista Indigenous Women: The Movement within the Movement." Lecture presented at the Department of Gender and Women's Studies, University of Washington, Seattle, 2011.

Garrison, Nanibaa. "Genomic Justice for Native Americans: Impact of the Havasupai Case on Genetic Research." *Science, Technology & Human Values* 38, no. 2 (2013): 201–23.

Geranios, Nickolas. "RezKast: A Sort of YouTube for Native Americans." *Cherokee Phoenix*, February 11, 2009. Accessed January 10, 2013. www.cherokeephoenix.org/20024/Article.aspx.

Goldberg-Hiller, Jonathon. "Borders of Kinship: Species/Race/Indigeneity." Workshop at the B/Ordering Violences Seminar at the Simpson Center for the Humanities, University of Washington, Seattle, May 24, 2013.

Gordon, Andrew, Margaret Gordon, and Jessica Dorr. "Native American Technology Access: The Gates Foundation in Four Corners." *Electronic Library* 21, no. 5 (2001): 428–34.

Government Accountability Office. *Tribal Internet Access: Increased Federal Coordination and Performance Measurement Needed.* By Mark Goldstein. GAO 16-504T. Washington, D.C.: GAO, 2016.

Graveline, Frye Jean. "Idle No More: Enough Is Enough!" *Canadian Social Work Review* 29, no. 2 (2012): 293–300.

Grey, Steve. "Navajo Technical College Takes Internet to Hogans." *Tribal College Journal of American Indian Higher Education* 18, no. 3 (2007): 39.

Guidotti-Hernandez, Nicole. *Unspeakable Violence: Remapping U.S. and Mexican National Imaginaries.* Durham, N.C.: Duke University Press, 2011.

Harmon, Amy. "Indian Tribe Wins Fight to Limit Research of Its DNA." *New York Times*, April 21, 2010.

Harris, LaDonna, Stephen Sachs, Benjamin Broome, and Jondodev Chaudhuri. "Returning to Harmony through the Wisdom of the People: Applying Traditional Principles to Develop Appropriate and Effective Indian Tribal Governance; Returning Indian Nations to Culturally Appropriate Forms of Decision Making." In *Re-creating the Circle: The Renewal of Indian Self-Determination*, edited by LaDonna Harris, Stephen Sachs, and Barbara Morris, 201–50. Albuquerque: University of New Mexico Press, 2011.

Havasupai Tribe of the Havasupai Reservation v. Arizona Board of Regents and Therese Ann Markow, Arizona Court of Appeals nos. 1 CA-CV 07-0454, 1 CA-CV 07-0801 (2008).

Hays, Robert. *Editorializing "The Indian Problem": The New York Times on Native Americans, 1860–1900.* Carbondale: Southern Illinois University Press, 2007.

Heidegger, Martin. *The Question Concerning Technology and Other Essays*. Edited and translated by William Lovitt. New York: Harper Colophon, 1977.

Heppler, Jason. "Framing Red Power: The American Indian Movement, the Trail of Broken Treaties, and the Politics of Media." PhD diss., University of Nebraska, Lincoln, 2009.

Hernandez Silva, Hector Cuauhtemoc. *Insurgencia y autonomía: Historia de los pueblos Yaquis, 1821–1910.* Mexico City: Instituto Nacional Indigenista, 1996.

Hicks, Sarah. "Intergovernmental Relationships: Expressions of Tribal Sovereignty." In *Rebuilding Native Nations: Strategies for Tribal Governance*, edited by Miriam Jorgensen, 246–72. Tucson: University of Arizona Press, 2007.

Holm, Tom, J. Diane Pearson, and Ben Chavis. "Peoplehood: A Model for the Extension of Sovereignty in American Indian Studies." *Wicazo Sa Review* 18, no. 1 (2003): 7–24.

Howard, Phil. *The Internet and Islam: The Digital Origins of Dictatorship and Democracy.* New York: Oxford University Press, 2010.

Howard, Phil, and Muzzamil Mohammed Hussain. *Democracy's Fourth Wave? Digital Media and the Arab Spring.* London: Oxford, 2013.

Howe, Craig. "Cyberspace Is No Place for Tribalism." *Wicazo Sa Review* 13, no. 2 (1998): 19–28.

———. "New Architecture on Indigenous Lands." Presentation at the Illinois School of Architecture, Urbana, April 2014.

HPWREN. "HPWREN's Continued Collaboration with Native Americans Results in Various Opportunities across Research, Education, and First Responder Activities." *HPWREN News*, July 1, 2005. Accessed January 10, 2013. http://hpwren.ucsd.edu/news/050701.html.

Hu-Dehart, Evelyn. *Missionaries, Miners, and Indians: History of Spanish Contact with the Yaqui Indians of Northwestern New Spain, 1533–1830*. Tucson: University of Arizona Press, 1981.

———. *Yaqui Resistance and Survival: Struggle for Land and Autonomy, 1821–1910*. Madison: University of Wisconsin Press, 1984.

Indian Self-Determination and Education Assistance Act of 1975 25 § 450.

Indigenous Commission for Communications Technologies in the Americas. *The Plan: Indigenous Peoples Empowering Themselves through Technology*. Ottawa: ICCTA, 2009.

"Internet Group Ranks Montana among the 'Disconnected Dozen.'" *Associated Press State and Local Wire*, July 29, 1999.

Jacoby, Karl. *Shadows at Dawn: An Apache Massacre and the Violence of History*. New York: Penguin Books, 2009.

John, Sonja. "Idle No More: Indigenous Activism and Feminism." *Theory in Action* 8, no. 4 (2015): 38–54.

Jorgensen, Miriam, Traci Morris, and Susan Feller. *Digital Inclusion in Native Communities: The Role of Tribal Libraries*. Oklahoma City, Okla.: Association of Tribal Archives, Libraries, and Museums, 2014.

Kauanui, J. Keuhalani. *Hawaiian Blood: Colonialism and the Politics of Sovereignty and Indigeneity*. Durham, N.C.: Duke University Press, 2008.

Kiefer, Michael. "Havasupai Ends Regents Lawsuit with Burial." *Arizona Republic*, April 22, 2010.

Kramer, Becky. "Face Time: Fast Horse Is Bridging Digital Divide." *Spokesman-Review*, April 4, 2011.

Krebs, Allison B. "Indigenous Information Ecology: Vanishing Indians Throwing Off Our Invisibility Cloaks Rushing into the 21st Century." Lecture presented at the School of Information Resources and Library Science, University of Arizona, Tucson, March 10, 2008.

———. "Native America's Twenty-First-Century Right to Know." *Archival Science* 12 (2012): 173–90.

Kroker, Arthur. *The Will to Technology and the Culture of Nihilism: Heidegger, Nietzsche, Marx*. Toronto: University of Toronto Press, 2004.

Kuhrt, Danielle. "Valerie Fast Horse." *Boise State University: Women Making History*, 2011. Accessed June 5, 2013. http://womenscenter.boisestate.edu/women/women-making-history/2011-honorees/valerie-fast-horse/.

LaDuke, Winona. *All Our Relations: Native Struggles for Land and Life*. London: South End Press, 1999.

Landzelius, Kyra. *Native on the Net: Indigenous and Diasporic Peoples in the Virtual Age*. London: Routledge, 2006.

Latour, Bruno. *Reassembling the Social: An Introduction to Actor-Network-Theory*. Oxford: Oxford University Press, 2007.

———. "Technology Is Society Made Durable." In *A Sociology of Monsters: Essays on Power, Technology, and Domination*, edited by John Law, 103–31. London: Routledge, 1991.

Law, John. "Power, Discretion, and Strategy." In *A Sociology of Monsters: Essays on Power, Technology, and Domination*, edited by John Law. New York: Routledge, 1991.

———, ed. *A Sociology of Monsters: Essays on Power, Technology, and Domination*. London: Routledge, 1991.

Linthicum, Leslie. "Historic Day in Shiprock." *Albuquerque Journal*, April 23, 2000.

Luisetti, Federico, John Pickles, and Wilson Kaiser, eds. *The Anomie of the Earth: Philosophy, Politics, and Autonomy in Europe and the Americas*. Durham, N.C.: Duke University Press, 2015.

Mander, Jerry. *The Failure of Technology and the Survival of the Indian Nations*. San Francisco: Sierra Club Books, 1991.

Martinez-Torres, Maria Elena. "Civil Society, the Internet, and the Zapatistas." *Peace Review* 13, no. 3 (2001): 347–55.

McMahon, Richard. "The Institutional Development of Indigenous Broadband Infrastructure in Canada and the United States: Two Paths to Digital 'Self-Determination.'" *Canadian Journal of Communication* 36 (2011): 115–40.

Medin, Douglas, and Megan Bang. *Who's Asking?: Native Science, Western Science, and Science Education*. Cambridge, Mass.: MIT Press, 2014.

Medina, José. *The Epistemology of Resistance: Oppression, Epistemic Injustice, and Resistant Imaginations*. Cambridge: Oxford University Press, 2012.

Merjian, Armen. "An Unbroken Chain of Injustice: The Dawes Act, Native American Trusts, and *Cobell v. Salazar*." *Gonzaga Law Review* 46 (2010): 609–716.

Meyer, Manu Aluli. *Ho'oulu: Our Time of Becoming; Hawaiian Epistemology and Early Writings*. Honolulu: Ai Pohaku Press, 2003.

Mezzadra, Sandro, and Brett Nielson. *Border as Method, or the Multiplication of Labor*. Durham, N.C.: Duke University Press, 2013.

Mignolo, Walter. "Anomie, Resurgences, and De-Noming." In *The Anomie of the Earth: Philosophy, Politics, and Autonomy in Europe and the Americas*, edited by Federico Luisetti, John Pickles, and Wilson Kaiser, vii–xv. Durham, N.C.: Duke University Press, 2015.

———. *The Darker Side of the Renaissance: Literacy, Territoriality, and Colonization*. Durham, N.C.: Duke University Press, 2003.

———. *Local Histories / Global Designs: Coloniality, Subaltern Knowledges, and Border Thinking*. Princeton, N.J.: Princeton University Press, 2012.

Miki Maaso, Felipe S. Molina, and Larry Evers. "The Elder's Truth: A Yaqui Sermon." *Journal of the Southwest* 35, no. 3 (1993): 225–317.

Miller, John C., and Christopher P. Guzelian. "The Spectrum Revolution: Deploying Ultrawideband Technology on Native American Lands." *11 Commlaw Conspectus* 277 (2003): 277–305.

"Mont. Stimulus-Funded Internet Expansion Assailed." *Associated Press State and Local Wire*, November 4, 2009.

Moraña, Mabel, Enrique Dussel, and Carlos A. Jáuregi, eds. *Coloniality at Large: Latin America and the Postcolonial Debate*. Durham, N.C.: Duke University Press, 2008.

Morris, Traci, and Sascha Meinrath. *New Media, Technology, and Internet Use in Indian Country: Quantitative and Qualitative Analyses*. Phoenix, Ariz.: Native Public Media; Washington, D.C.: New America Foundation, 2009.

Moses, L. G. *Wild West Shows and the Images of American Indians: 1883–1933*. Albuquerque: University of New Mexico Press, 1996.

Mumford, Lewis. *Technics and Civilization*. Chicago: The University of Chicago Press, 1934.

Munster, Anna. *An Aesthesia of Networks: Conjunctive Experience in Art and Technology*. Cambridge, Mass.: MIT Press, 2013.

Nabakov, Peter. *Where Lightning Strikes: The Lives of American Indian Sacred Places*. New York: Penguin, 2007.

Nahon, Karine, and Jeff Helmsley. *Going Viral*. Cambridge: Polity, 2013.
Nardi, Bonnie, and Vicki O'Day. *Information Ecologies: Using Technology with Heart*. Cambridge, Mass.: MIT Press, 2000.
National Telecommunications and Information Administration and Federal Communications Commission. "National Broadband Map (2013)." Accessed June 5, 2013. www.broadbandmap.gov.
Native American Broadband Association. "Native American Broadband Association, 2010." Accessed January 15, 2011. www.nativeamericanbroadband.org.
Navajo Nation Telecommunications Regulatory Commission. "History." 2012. Accessed December 14, 2012. www.nntrc.org/content.asp?CustComKey=21364&CategoryKey=21413&pn=Page&DomName=nntrc.org.
Navajo Technical College. "Internet to the Hogan." 2013. Accessed January 11, 2013. www.navajotech.edu/index.php/ith.
Niezen, Ron. *The Origins of Indigenism: Human Rights and the Politics of Identity*. Berkeley: University of California Press, 2003.
Noori, Margaret. "Waasechibiiwaabikoonsing Nd'anami'aami, 'Praying through a Wired Window': Using Technology to Teach Anishinaabemowin." *Studies in American Indian Literatures* 23, no. 2 (2011): 1–23.
Norrell, Brenda. "Yaqui Vicam Pueblo International Gathering for the Defense of Water." *Censored News*, November 21, 2012. Accessed August 3, 2016. http://bsnorrell.blogspot.com/2012/11/yaqui-vicam-pueblo-international.html.
Oakes, Leslie, and Joni Young. "Reconciling Conflict: The Role of Accounting in the American Indian Trust Fund Debacle." *Critical Perspectives on Accounting* 21, no. 1 (2010): 63–75.
Office of Technology Assessment, US Congress. *Telecommunications Technology and Native Americans: Opportunities and Challenges*. OTA-ITC-621. Washington, D.C.: US Government Printing Office, 1995.
Oliphant v. Suquamish 435 U.S. 191 (1978).
Olokotun, Deji. "Why Is a Tiny Island Nation Facing an Internet Shutdown?" *Access: Mobilizing for Global Digital Freedom*, May 14, 2015. Accessed September 3, 2015. www.accessnow.org/blog/2015/05/14/why-is-a-tiny-island-nation-facing-an-internet-shutdown.
Orticio, Gino. *Indigenous/Digital Heterogeneities: An Actor-Network-Theory Approach*. Doctoral thesis, Queensland University of Technology, 2013.
Pacific Railroad Acts, 12 § 489 (1862, 1863, 1864).
Painter, Muriel. *With Good Heart: Yaqui Ceremonies and Beliefs in Pascua Village*. Tucson: University of Arizona Press, 1986.
Palmer, Mark. "Cartographic Encounters at the Bureau of Indian Affairs Geographic Information System Center of Calculation." *American Indian Culture and Research Journal* 36, no. 2 (2012): 75–102.
Prashad, Vijay. *The Darker Nations: A People's History of the Third World*. New York: The New Press, 2008.
Pyrillis, Rita. "IT across the Navajo Nation," *FedTech Magazine*, 2009. Accessed December 12, 2012. http://fedtechmagazine.com/article/2010/01/it-across-navajo-nation.
Quijano, Anibal. "Colonialidad del poder y clasificacion social." *Journal of World-Systems Research* 11, no. 2 (2000): 342–87.
———. "Coloniality and Modernity/Rationality." In *Globalizations and Modernities*, edited by Goran Therborn, 41–51. Stockholm: Forksningsradnamnden, 1992.
———. "Coloniality of Power, Eurocentrism, and Latin America." *Nepantla: Views from the South* 1, no. 3 (2000): 549–54.

Raibmon, Paige. "Unmaking Native Space: A Genealogy of Indian Policy, Settler Practice, and the Microtechniques of Dispossession." In *The Power of Promises: Rethinking Indian Treaties in the Pacific Northwest*, edited by Alexandra Harmon, 56–85. Seattle: University of Washington Press, 2008.

Rantanen, Matt. "Tribal Digital Village: Southern California Tribal Chairmen's Association." Lecture presented at the Benton Foundation Broadband and Economic Development Summit, Dallas, April 2011.

Red Spectrum Communications, Inc. "Coeur d'Alene Tribe's Fiber to the Home Project (2013)." Accessed April 12, 2013. http://redspectrum.com/fibermain.html.

Richards, Thomas. *The Imperial Archive: Knowledge and the Fantasy of Empire*. London: Verso, 1996.

Richardson, Jayson, and Scott McLeod. "Technology Leadership in Native American Schools." *Journal of Research in Rural Education* 26, no. 7 (2011): 1–14.

Riley, Linda Ann, Bahram Nassarsharif, and John Mullen. *Assessment of Technology Infrastructure in Native Communities*. Economic Development Administration. Las Cruces: New Mexico State University, 1999.

Romero, Channette. "Envisioning a 'Network of Tribal Coalitions': Leslie Marmon Silko's *Almanac of the Dead*." *American Indian Quarterly* 26, no. 4 (2002): 623–40.

Roy, Loriene. "Four Directions: An Indigenous Educational Model." *Wicazo Sa Review* 13, no. 2 (1998): 59–69.

Russo, Tim. "Zapatista March: The Deafening Silence of Resurgence." *Upside Down World: Covering Activism and Politics in Latin America*, December 22, 2012. Accessed August 3, 2016. http://upsidedownworld.org/main/mexico-archives-79/4041-zapatista-march-the-defeaning-silence-of-resurgence.

Sandvig, Christian. "Connection at Ewiiaapaayp Mountain: Indigenous Internet Infrastructure." In *Race after the Internet*, edited by Lisa Nakamura and Peter Chow-White, 168–200. New York: Routledge, 2011.

Savard, Jean-François. "A Theoretical Debate on the Social and Political Implications of the Internet for the Inuit of Nunavut." *Wicazo Sa Review* 13, no. 2 (1998): 83–97.

Shaffer, Mark. "Havasupai Blood Samples Misused." *Indian Country Today*, March 9, 2004.

Shirley, Sheryl. "Zapatista Organizing in Cyberspace: Winning Hearts and Minds?" Paper presented at the Conference of the Latin American Studies Association, Washington, D.C., September 2001.

Shorter, David Delgado. "Hunting for History in Potam Pueblo: A Yoeme (Yaqui) Indian Deer Dancing Epistemology." *Folklore* 118 (2007): 283–307.

Silko, Leslie Marmon. *Almanac of the Dead*. New York: Penguin, 1994.

Silven, Eva. "Contested Sami Heritage: Drums and Sieidis on the Move." Paper presented at "EuNaMus 2012: Identity Politics, the Past, and the European Citizen," Brussels, January 26–27, 2012.

Simpson, Audra. "Mapping Sovereignty: Indigenous Borderlands." Paper presented at "B/ordering Violence: Boundaries, Gender, Indigeneity in the Americas: 2012–13 John E. Sawyer Seminar in Comparative Cultures," Simpson Center for the Humanities, University of Washington, Seattle, April 11, 2013.

———. *Mohawk Interruptus: Political Life across the Borders of Settler States*. Durham, N.C.: Duke University Press, 2014.

Sirois, John, Margaret Gordon, and Andrew Gordon. *Native American Access to Technology Program: A Progress Report*. Seattle: Evans School of Public Affairs, University of Washington, 2001.

Slotkin, Richard. *Regeneration through Violence: The Mythology of the American Frontier, 1600–1860*. Oklahoma City: University of Oklahoma Press, 2000.

Smith, Linda Tuhiwai. *Decolonizing Methodologies: Research and Indigenous Peoples*. London: Zed Books, 1999.

Southern California Tribal Digital Village. "CUWiN Announces Mesa Grande Reservation Installation." June 19, 2006. Tribal Digital Village. Accessed January 10, 2013. http://sctdv.net/node/101.

Spicer, Edward. *Cycles of Conquest: The Impact of Spain, Mexico, and the United States on Indians of the Southwest, 1533–1960*. Tucson: University of Arizona Press, 1967.

Srinivasan, Ramesh. "Tribal Peace: Preserving the Cultural Heritage of Dispersed Native American Communities." Paper presented at the International Conference on Cultural Heritage and Informatics, Berlin, September 2004.

Star, Susan Leigh, and Anselm Strauss. "Layers of Silence, Arenas of Voice: The Ecology of Visible and Invisible Work." *Computer Supported Cooperative Work* 8, nos. 1/2 (1999): 9–30.

Stevens, James. "E-Socials: Cultural Collaboration in the Age of the Electronic Inter-Tribal." In *Sovereign Bones: New Native American Writing*, edited by Eric Gansworth, 275–80. New York: Nation Books, 2007.

Stewart-Harawira, Makere. *The New Imperial Order: Indigenous Responses to Globalization*. London: Zed Books, 2005.

Subcomandante Insurgente Marcos. "The First Other Winds." In *The Speed of Dreams: Selected Writings, 2001–2007*, edited by Marco Canek Peña-Vargas and Greg Ruggiero, 302–17. San Francisco: City Lights, 2007.

———. "The Hand That Dreams When It Writes." In *The Speed of Dreams: Selected Writings, 2001–2007*, edited by Marco Canek Peña-Vargas and Greg Ruggiero, 140–43. San Francisco: City Lights, 2007.

———. "Other Intellectuals." In *The Speed of Dreams: Selected Writings, 2001–2007*, edited by Marco Canek Peña-Vargas and Greg Ruggiero, 337–46. San Francisco: City Lights, 2007.

Subcommittee on Management, Integration and Oversight of the Committee on Homeland Security, House of Representatives. *The Secure Border Initiative: Ensuring Effective Implementation and Financial Accountability of SBInet: Hearing before the Subcommittee on Management, Integration, and Oversight of the Committee on Homeland Security*. 109th Cong., 2nd sess., November 15, 2006.

Sullivan, Shannon, and Nancy Tuana, eds. *Race and Epistemologies of Ignorance*. New York: SUNY Press, 2007.

Tagaban, Brian. "Understanding the Current Regulatory Climate." Presentation at Tribal Telecom, Gila River, Arizona, February 2012.

Tahy, Emory, Nicolet Deschine Parkhurst, Traci Morris, and Karen Mossberger. "The Digital Reality: E-Government and Access to Technology and Broadband for American Indian and Alaska Native Populations." Presentation at the Sixteenth Annual International Conference on Digital Government Research, Phoenix, Arizona, 2015.

Tehranian, Majid. *Global Communication and World Politics: Domination, Development and Discourse*. Boulder, Colo.: Lynne Reiner, 1999.

Telecommunications Act of 1996 Pub. LA. 104, 110 Stat. 56 (1996).

"Treaty with the Sioux-Brule, Oglala, Miniconjou, Yanktonai, Hunkpapa, Blackfeet, Cuthead, Two Kettle, San Arcs, and Santee-and-Arapaho," April 29, 1868. *General Records of the United States Government*, Record Group 11, National Archives.

Tuck, Eve, and Wayne Yang. "Decolonization Is Not a Metaphor." *Decolonization: Indigeneity, Education, and Society* 1, no. 1 (2012): 1–40.

Tully, James. *Public Philosophy in a New Key.* New York: Cambridge University Press, 2008.
Tupper, Jennifer. "Social Media and the Idle No More Movement: Citizenship, Activism, and Dissent in Canada." *Journal of Social Science Education* 13, no. 4 (2014): 87–94.
Turner, Frederick Jackson. "The Significance of the Frontier in American History." Paper presented at the World's Columbian Exposition, Chicago, 1893. Washington, D.C.: American Historical Association, 1893.
Two Horses, Michael. "Gathering around the Electronic Fire: Persistence and Resistance in Electronic Formats." *Wicazo Sa Review* 13, no. 2 (1998): 29–43.
United Nations General Assembly. "Article 20: The Promotion, Protection, and Enjoyment of Human Rights on the Internet." Universal Declaration of Human Rights, 2012.
———. United Nations Declaration on the Rights of Indigenous Peoples: Resolution / Adopted by the General Assembly. October 2, 2007, A/RES/61/295.
United States. "Treaty with the Sioux-Brule, Oglala, Miniconjou, Yanktonai, Hunkpapa, Blackfeet, Cuthead, Two Kettle, San Arcs, and Santee-and-Arapaho," April 29, 1868. *General Records of the United States Government*, Record Group 11, National Archives.
United States Congress. Native American Telecommunications Act of 1997. 105th Cong., 1997–98. H.R. 486.
United States Customs and Border Patrol. *US Customs and Border Protection Performance and Accountability Report, Fiscal Year 2008.* Washington, DC: Office of Finance, 2008.
United States Government Accountability Office. *Alien Smuggling: DHS Needs to Better Leverage Investigative Resources and Measure Program Performance Along the Southwest Border: Report to Congressional Requesters.* GAO-10-2328, 2010.
———. *Secure Border Initiative: DHS Needs to Reconsider Its Proposed Investment in Key Technology Program.* GAO-10-340, 2010.
Unrau, William. *Tending the Talking Wire: A Buck Soldier's View of Indian Country, 1863–1866.* Salt Lake City: University of Utah Press, 1979.
Upper One Games. "Never Alone / Kisima Innitchuna." Accessed September 3, 2015. http://neveralonegame.com.
von Baeyer, Hans. *Information: The New Language of Science.* Cambridge, Mass.: Harvard University Press, 2004.
Walia, Harsha. *Undoing Border Imperialism.* Oakland, Calif.: AK Press, 2013.
Wallerstein, Immanuel. *The Politics of the World-Economy: The States, the Movements and the Civilizations.* Cambridge: Cambridge University Press, 1984.
Watts, Vanessa. "Indigenous Place-Thought and Agency amongst Humans and Non-humans (First Woman and Sky Woman Go on a European World Tour!)." *Decolonization: Indigeneity, Education, and Society* 2, no. 1 (2013): 20–34.
Wilkins, David. *American Indian Sovereignty and the U.S. Supreme Court: The Masking of Justice.* Austin: University of Texas Press, 1997.
Williams, J. D. *Statement by J. D. Williams, Cheyenne River Sioux Telephone Authority, before the Federal Communications Commission Regarding Overcoming Obstacles to Telephone Service to Indians on Reservations*, March 23, 1999.
Wilson, Pamela, and Michelle Stewart. *Global Indigenous Media: Cultures, Poetics, and Politics.* Durham, N.C.: Duke University Press, 2008.
Wilson, Shawn. *Research Is Ceremony.* Black Point, Nova Scotia: Fernwood Publishing, 2009.
Worcester v. Georgia 31 US (6 Pet.) 515 (1832).
"Yaquis mantienen bloqueada una carretera de Sonora." *Diario de Yucatán*, December 27, 2012. Accessed August 3, 2016. http://yucatan.com.mx/mexico/yaquis-mantienen-bloqueada-una-carretera-de-sonora.

INDEX

Charlotte Coté and Coll Thrush
Series Editors

Indigenous Confluences publishes innovative works that use decolonizing perspectives and transnational approaches to explore the experiences of Indigenous peoples across North America, with a special emphasis on the Pacific Coast.

A Chemehuevi Song: The Resilience of a Southern Paiute Tribe
by Clifford E. Trafzer

Education at the Edge of Empire: Negotiating Pueblo Identity in New Mexico's Indian Boarding Schools
by John R. Gram

Indian Blood: HIV and Colonial Trauma in San Francisco's Two-Spirit Community
by Andrew J. Jolivette

Native Students at Work: American Indian Labor and Sherman Institute's Outing Program, 1900–1945
by Kevin Whalen

California through Native Eyes: Reclaiming History
by William J. Bauer Jr.

Unlikely Alliances: Native Nations and White Communities Join to Defend Rural Lands
by Zoltán Grossman

Dismembered: Native Disenrollment and the Battle for Human Rights
by David E. Wilkins and Shelly Hulse Wilkins

Network Sovereignty: Building the Internet across Indian Country
by Marisa Elena Duarte

www.ingramcontent.com/pod-product-compliance
Ingram Content Group UK Ltd.
Pitfield, Milton Keynes, MK11 3LW, UK
UKHW040802120225
454975UK00002B/41